THE Happy Atom STORY 3

READ
A Fantasy Tale
LEARN
Basic
Chemistry

BOOK 3

IRENE P. REISINGER
ILLUSTRATIONS BY SARA K. WHITE
GRAPHICS BY IRENE P. REISINGER

Archway Publishing books may be ordered through booksellers or by contacting:

Archway Publishing
1663 Liberty Drive
Bloomington, IN 47403
www.archwaypublishing.com
844-669-3957

ISBN: 978-1-4808-9196-8 (sc)
ISBN: 978-1-4808-9195-1 (hc)
ISBN: 978-1-4808-9197-5 (e)

Print information available on the last page.

Archway Publishing rev. date: 12/29/2020

The Happy Atom Story

A Four Book Series

Book 1 Guy Learns About the Periodic Table

Guy meets Wish Star on the mountain and is introduced to chemistry's unique vocabulary. Through the magic tunnel in Professor Terry's lab, Guy enters Periodic Table Land and is taught to interpret the Periodic Table to find information about the atom and the particles in the atom. This begins his understanding of Chemistry,

Book 2 Guy's Adventures With the Elements

Using the Periodic Table, Guy draws Bohr models of the atoms. Flying around Periodic Table Land, he meets the elements in each of the Chemical Families who will be part of his chemistry experience. In the end Sodium, the scientist of Periodic Table Land, discovers the secret of becoming a Happy Atom as he visits the Noble Gases on 8th Street. That secret is really how compounds are formed.

Book 3 Guy Learns to Write Chemical Formulas

Guy learns how compounds are formed watching the sad little elements become Happy Atoms. As Guy watches the compounds forming from the invisible bubble, he learns the meaning of chemical formulas. Finally he learns to construct chemical formulas using the Valence Method.

Book 4 Guy Learns to Balance Equations

Polyatomic Ions enter Periodic Table Land as Superhero Counselors to help the Chemical Families whose members feel misunderstood. Then the Polyatomic Ions learn to form compounds and Guy uses the Valence method to write their formulas. After that Guy learns about Chemical Reactions and writes their equations. The little elements with seesaws pop in to teach Guy to balance equations. In the end there is a Chemistry Parade as a review of all the principles of chemistry that Guy learned with Professor Terry's help and the help of all the dear little elements in Periodic Table Land

DEDICATION

TO

Sister Regina Mercedes, CSJ

My Chemistry Teacher
St. Brendan's High School
Brooklyn, New York

and

John Carlin Ph.D.

My Inorganic Chemistry Professor
Fordham University
New York City

Thank you for helping me understand Chemistry.

TO

MY LOVING PARENTS

Irene and John Murray

Thank you for my happy childhood.

TO

My wonderful husband, **Fred** and my five loving children
Terry, Mary, Kathy, Freddy and John
With All My Love

THE INTRODUCTION

A Fantasy Tale Teaches Basic Chemistry

If you have read BOOK 1 and 2 of *The Happy Atom Story,* you know that the principles of chemistry are woven into a fantasy tale for two reasons. First and foremost, the story makes the rather difficult principles of chemistry more understandable. Secondly, the story captures and holds the attention of young readers long enough to have them learn the principle of chemistry being explained. It is also a proven fact that it is easier to remember a story than a list of facts. As the student recalls the story the chemical principles woven into the story are remembered as well.

The story originated while I was teaching a class of 8th graders how compounds are formed by sharing or exchanging electrons. It was right after lunch and the A/C was not working. Getting their attention seemed a remote possibility. I knew if I didn't do something quickly, they were not going to learn compound formation, and the rest of the unit depended on understanding this. When I gave up the traditional way of explaining compound formation, and told the class a story about sad little atoms that learned to be happy, I had the attention of every student in my class. At the end of the lesson that day, every single student had learned that these sad little atoms became happy when they shared or exchanged electrons. When the atoms became happy they became a beautiful new creation called a compound. This story about compound formation was so effective I continued using it for the next 18 years.

Again if you have read Books 1 and 2, you know *The Happy Atom Story* is about a boy named Guy vacationing on the mountain for the summer with his family. He loves the stars, planets and nature all around him and is in awe of his magnificent world. When he wishes on a star, Wish Star appears beside him and tells Guy that chemistry will teach him about a whole new world hidden under the surface of the visible world he loves.

Wish Star begins Guy's introduction to the world of chemistry, but he can't stay all summer. So, he introduces Guy to Professor Terry. This is when Guy's real adventures begin. Professor Terry has a magic Periodic Table in her lab that has an entrance into the fantasy world of Periodic Table Land. Guy's adventures take place in this mystical world. It's there that the elements, the silly electrons, the proper protons and all the little atoms eagerly share their knowledge of the world of chemistry with Guy. The reader learns chemistry along with Guy. When BOOK 1 ends, Guy knows how to interpret the Periodic Table to help him understand the structure of the atom.

In Book 2, Guy learns to draw the structure of atoms for the elements Atomic Numbers 1 through 20. Then Guy meets the Chemical Families and gets to know something about each of the elements in these families. Most important he learns that elements in the same family that appear so very different are alike because they all have the same number of electrons in their Outside Energy Level. That's why they are in the same Chemical Family. Meeting the different Chemical Families and the elements in these families is necessary to understand how compounds are formed in Book 3 that you are about to read.

Book 2's final adventure ends when Sodium discovers how elements can become compounds by becoming Happy Atoms. It's all about the element's Outside Energy Levels. This is the information needed to understand compound formation, the subject of Book 3

Book 3 is about compounds and constructing chemical formulas. This is what you are about to enjoy. Read Book 3 and understand how compounds are formed. Chemistry is not difficult, it's just finding the right medium to convey the principles of chemistry. This book does just that. The Happy Atom Story was used successfully to teach chemistry to 8th Grade students for 18 years.

The Happy Atom Story was a success in helping students learn chemistry, and my colleagues never gave up urging me to publish it. Those in the Special Education Department amazed that their students were understanding chemistry and constructing chemical formulas, insisted that I should publish *The Happy Atom Story*. Barbara Brooks who worked with these special students said often in her very quiet way, "If you publish your story, so many students will be helped." My middle school students loved the story and excitedly added touches to it over the years.

One day by chance, I ran into a student I had taught in middle school. At the time of our meeting .she was then a senior in college. She told me how my story affected her life. "In freshman year, I decided to drop chemistry and give up my dream of becoming a doctor. I went home and thought, why can't my professor make chemistry easy like Mrs. Reisinger." She said, "I thought about how you used happy atoms to explain chemistry. Through your story I was able to figure out what my 'impossible to understand Professor was trying to say. I passed that course, stayed in the Pre-Med program. Last week I received my acceptance into medical school. Thank you, Mrs. Reisinger."

That student's story taught me that my Happy Atom Story had a value beyond the middle school classroom. I knew reading my story would help teachers by providing them with a different approach to reach more middle school children. It would also help more students succeed in high school chemistry. That student's story made me realize that reading my book, *The Happy Atom Story,* could help students at any level. *The Happy Atom Story* provides an overview of basic chemistry, giving students a chance to better understand chemistry when taking higher level classes. From day one they have a point of reference to see where principles being explained in more technical language fit into the big picture of chemistry.

My book, *The Happy Atom Story,* can be an asset to teachers who find students having difficulty understanding certain principles of chemistry. It provides a fresh approach to the difficult principles that students just don't seem to get when explained in the traditional way.

About the Author

The author was able to write this book because she has a thorough understanding of basic chemistry, the middle school child, and an imagination. Her education plus her life experiences prepared her to write this story. Her background in chemistry gave her the ability to turn the principles of chemistry into a story without distorting their meaning. She received the Gold Medal for highest honors in chemistry at graduation from Fordham University where she earned her undergraduate degree with a double major: Chemistry and Education. After graduation, she passed the New York City exam to become a licensed chemist. She worked as the chemist with a team of doctors doing research on the kidney for two years at New York University Medical Center in the Department of Renal Physiology. During this time she earned her Masters Degree in Science Education from New York University, Washington Square. The emphasis there was on creativity. Next she taught science for a short time in a junior high school in Brooklyn before marrying her college sweetheart who had now become an officer in the US Marine Corps.

The next part of her life contributed to the story-telling ability that was needed to create *The Happy Atom Story.* Her marine husband was gone six months to a year at a time giving her the opportunity to be creative raising her five little Reisingers. At one point she created a kindergarten in her home for Kathy, her third child, and nine other neighborhood five year olds. Then, while her husband worked at NSA and all her children were in school, she taught for five years in the Primary Department of Trinity Lower School in Howard County, Maryland. There she taught second graders how to write three paragraph compositions motivated by pictures. Her science background was utilized as she enriched the science program for the Primary grades. All this definitely played a part in developing her imagination.

At that time her friends who were chemistry majors with her in college were teaching at medical schools and universities. She wondered what she was doing still involved with young children. Yet she loved what she was doing. Little did she know that she was being groomed to one day successfully teach chemistry to middle school students and eventually author this book. It is amazing how life experiences all blend together to be the perfect preparation for some future achievement as yet unknown. Her involvement with young children was preparing her to later be able to explain the difficult principles of chemistry on a level young students would understand.

When her husband retired she began her career teaching middle school. She taught a course whose objective was to prepare middle school students to succeed in high school chemistry. She taught the middle school students basic chemistry for 19 years. It was in that 8th grade classroom that she discovered an extremely effective way to maintain the attention of middle school students and at the same time have them learn chemistry. This was accomplished by weaving the principles of chemistry into a story that helped the students learn chemistry. That effective story is the basis of this book, *The Happy Atom Story.* Without her life experiences and her knowledge of Chemistry this would never have happened.

Editing for Science Integrity

The background of the author is presented to assure the reader that the book contains valid science. The author's degree in chemistry provided her with far more knowledge than she needed to teach basic chemistry. For her undergraduate degree she had Inorganic, Organic and Physical Chemistry. She also had Qualitative and Quantitative Analysis which prepared her to be a chemist. But to assure you that the principles as explained in this book are accurate, she invited two chemists to read and critique her manuscript. It passed their inspection. The backgrounds of these chemists are described next.

Joyce M. Donohue, Ph.D.—taught chemistry on every level: high school, community college, university and graduate school. Presently, she's working for the EPA doing long term studies on water pollutants. In addition she taught chemistry at a local community college. Having taught high school Chemistry, Joyce's final comment was, "High school teachers will really appreciate the way your book prepares their students."

Mr. Robert L. Zimmerman, Jr., M.S.—is a chemist who was the Director of the Research Lab for Customs and Border Protection, a part of Homeland Security, until he retired. He is now working for the American Association for Laboratory Accreditation to link laboratories throughout the country to the *International Standard for Chemistry Labs*. After reading *The Happy Atom Story,* he especially liked the study suggestions the book provides.

I appreciated these professionals taking the time to read my manuscript assuring you that you are reading solid basic chemistry. It is important that a reader knows from the outset that the science they are reading is valid.

I would like also to recognize Teresa Halsema and Cathy Turner for editing all four books of the Happy Atom Story. Cathy before retiring was head of the English Department at Woodbridge Senior High School, Woodbridge VA. Thank you both for your long hours of work.

Groups That Should Consider Buying *The Happy Atom Story*

Parents

My neighbor is a parent who wants to see her child succeed. When I told her about the Happy Atom story she said, "Write that story in a hurry. My daughter is taking AP Chemistry in the fall. My husband and I do not remember enough chemistry to help." So she pointed out that this kind of a book would be eagerly received by parents who wish to help their child in such a challenging subject.

My granddaughter Kristine's husband, Phil, has a Ph.D. in chemistry and works as a research chemist. He is interested in getting hold of my book and using it to introduce chemistry to his three little children. He plans to read parts of it to them before they learn to read, letting them enjoy the pictures as he tells the story.

The family across the street has a child who is an avid reader. Both his parents are physicians They want my book to encourage this son to extend his interest in science to include chemistry.

Home School Groups

There are members of several home school communities in my area who expressed an interest in my book. They would like to use it for students at a young age to help them become grounded in the basics of chemistry before the students take high school chemistry. It can also help teachers get another perspective on chemistry, especially if they were not chemistry majors in college. I've spoken to many other parents who home school their children, and they see a great need for this kind of a book in their home school communities.

Students at Any Level

Students at any level—middle school, high school, community college or even the university would do well to read *The Happy Atom Story* during the summer before taking chemistry. The book provides an understandable framework on which to hang the many more technical concepts provided at higher levels. All students will benefit from looking at chemistry from a new perspective.

Special Education Teachers

There are requirements presently for special education teachers to demonstrate how they are crafting their lessons to accommodate a variety of learning styles. *The Happy Atom Story* definitely does this in the field of chemistry. The reader will appreciate how this book makes theoretical subject matter visual and more understandable. Also the stories help students understand complex concepts with explanations through stories that help students remember the chemical principles longer.

Teachers with Chemistry in their Curriculum

Many 6th Grade teachers have a unit on chemistry in their curriculum. *The Happy Atom Story* brings chemistry down to a level their students can understand. For higher level chemistry teachers it affords a fresh perspective on how to explain principles to students who are having trouble understanding chemistry because of the complex vocabulary in high school and college level text books.

College Students with Chemistry as a Requirement

Many college students, who are not chemistry majors, are required to take chemistry for their degree. This is often a challenge. By reading the book they will be establishing a base to build on before they take the higher level chemistry course.

Students studying for a degree in Science Education would especially profit from reading *The Happy Atom Story.* This book will be a good resource if they some day have to teach chemistry as part of a science class.

The General Public

Educators hope that the general public will increase their knowledge of science so they can have a better understanding of how science affects their lives. Research has shown that chemistry is the least understood of all the sciences. Reading *The Happy Atom Story* will give the reader an overview of what is basic to chemistry. I've spoken to many obviously intelligent people at different events and most of them verbalized that they had a hard time understanding chemistry when they took the class in high school. Others said, "I flat out failed that subject." One said, "I took the course and got a good grade, but I never did figure out what it was all about. I just memorized everything." Doctors, I've talked with agreed that chemistry is a difficult subject, and a book like mine is seriously needed. One podiatrist said, "I'd read it just for the heck of it."

Foreign Countries

The internet indicates foreign countries are looking for science books written in lay man's language. *The Happy Atom Story* fills this need in the area of basic chemistry.

BOOK 3

COMPOUND FORMATION

Guy Learns How Compounds Are Formed and How to Write Chemical Formulas

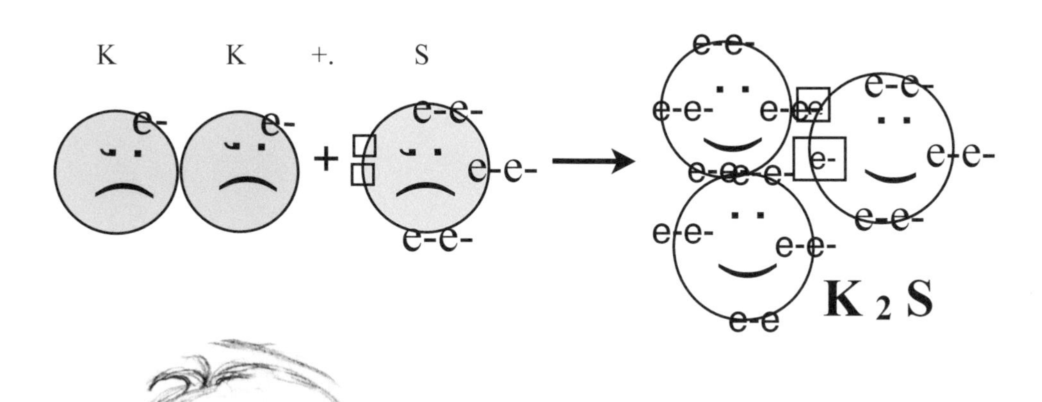

BOOK 3
TABLE OF CONTENTS

CHAPTER	THE ELEMENTS BECOME HAPPY ATOMS OR HOW COMPOUNDS ARE FORMED	PAGES
1	The Alkali Metals and the Halogens Experiment	14 - 25
2	The Alkali Metals and the Halogen Family	26 - 35
3	The Alkaline Earth Metals and the Oxygen Family	36 - 45
4	The Boron Family and The Nitrogen Family	46 - 54
5	The Alkali Metals and The Oxygen Family	55 - 61
6	The Alkali Metals and The Nitrogen Family	62 - 68
7	The Alkaline Earth Metals and The Halogens	69 - 75
8	The Boron Family and The Halogens	76 -84
9	The Alkaline Earth Metals and The Nitrogen Family	85 - 94
10	The Boron Family and The Oxygen Family	95 - 102
11	The Valence Method——Writing Chemical Formulas	103-117

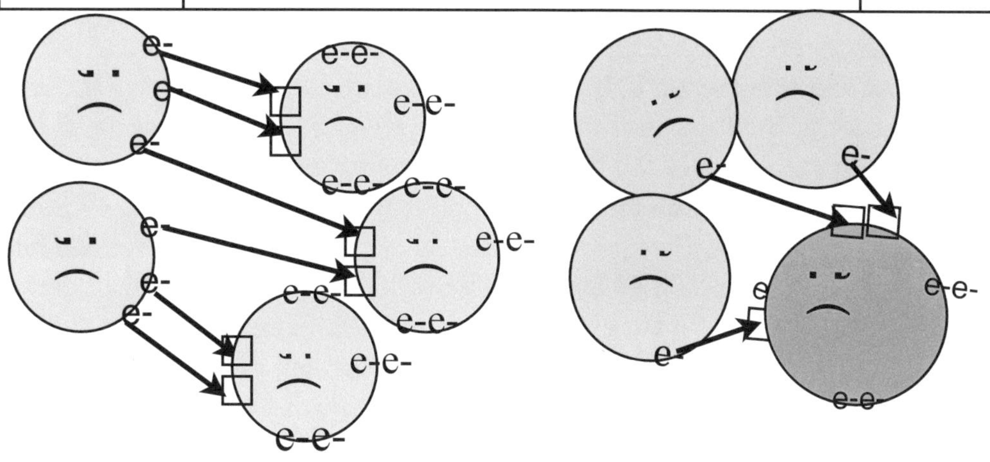

Can the Alkali Metals and the Halogens Become Happy Atoms?

Chapter 1

Morning came. Guy woke up extra early anxious for the day to start. Today was the day he was going back to Periodic Table Land to see if Sodium could really create a compound with Chlorine and become a Happy Atom like the elements on 8th Street. He jumped out of bed, pulled on his clothes and joined his parents for a good breakfast in the cabin's cozy kitchen. Then Guy rushed down the mountain hardly taking time to notice the little chipmunks scurrying in and out of the rocks on the side of the road. When he reached the entrance to Professor Terry's lab, he eased open the door and peered into the lab. There was Professor Terry waiting for him back by the Periodic Table.

"Hi Guy! Glad you got here early. The sooner we leave for Periodic Table Land, the sooner we'll get to Sodium's place, and your new adventure will begin." Zigzagging through the maze of slate topped lab tables, Guy made his way back to join Professor Terry. Together they leaned on the Periodic Table. The whole section of the wall pushed in and flipped around. Bells, whistles and sirens screamed——adventure ahead.

Professor Terry and Guy began sliding down the tunnel of mirrors reflecting a thousand sparkling lights. Magical sprinkle dust enveloped them like a silky blanket. They heard soft comforting background music. Then, in an instant, they were in that strange mysterious world, Periodic Table Land. Guy was as amazed as ever at the sight of those colorful blinking lights surrounding the street signs, the adorable houses of an amazing variety of shapes and sizes, and the dear little elements running around here, there and everywhere. Guy never ceased being thrilled that he was actually visiting this fantasy world.

Professor Terry said, "Look how happy you are to be back here!"

Guy said, "I love the street signs surrounded by those colorful blinking lights. Now I even know why those lights are a special color. The lights are the chemical family's special color and each street is the home of a different chemical family. I now know the names of the chemical families that live on each of the streets. The green blinking lights on 1st Street mark the home of the Alkali Metals, Group 1A on the Periodic Table. The amber street lights on 2nd Street mark the home of the Alkaline Earth Metals, Group 2A on the Periodic Table. B Avenue with all the B Groups is the home of the Transition Metals. After that the light gray blinking lights mark 3rd Street, the home of the Boron Family, Group 3A on the Periodic Table; the black lights mark 4th Street, the home of the Carbon Family, Group 4A; 5th Street's red blinking lights mark the home of the Nitrogen Family, Group 5A; 6th Street"s yellow blinking lights mark the home of the Oxygen Family, Group 6A; 7th Street's yellowish green blinking lights mark the home of the Halogens, Group 7A on the Periodic Table. At the far end of Periodic Table Land is Group 8A, 8th Street, the home of the Noble Gas Family. I love their purple blinking lights"

Professor Terry added, “Remember that all the element houses that you see here are what were the element boxes on the Periodic Table. The address of each element’s house is his Atomic Number. Sodium’s address is #11, 1st Street. On the Periodic Table Sodium’s element box would be in Group 1A and his Atomic Number 11. Check this out on your Periodic Table.”

Guy sat down, opened his backpack and pulled out his Periodic Table. He said, “Now all I have to do is find Group 1A and slide my finger down to find Sodium. There it is, and Sodium’s Atomic Number is 11. Guess you were right. The element’s house number here in Periodic Table Land is the Atomic Number. The street is the Group Number.” Guy tucked the important paper file into his backpack and they moved on.

Then Guy glanced down the left side of town where there were many different colored lights on short poles with numbers on them. He recalled, “They are the Period numbers on my Periodic Table showing how many energy levels elements have. Sodium is in Period 3. So his atom has 3 energy levels.” Guy checked Sodium’s Bohr model in Book 2. Indeed Sodium had 3 energy levels. Guy added, “All the elements in the Period 3 row, all the way over to Group 8A at the other end of the Periodic Table have 3 energy levels too.”

Professor Terry stood there as Guy gazed in appreciation of the unique houses that were once the element boxes on the Periodic Table. “Every time we come to Periodic Table Land I’m fascinated by the elements’ unique houses. I really like that tree house over there. Look there’s an element climbing up the ladder to the tree house right now.”

Professor Terry answered, “There’s a lot to see here and lots of magic in Periodic Table Land. There’s even some new magic that you haven’t seen yet. Let’s get to Sodium’s house and see what special surprise he has for us today.”

They passed Lithium’s place on 1st Street and soon they were at Sodium’s house. Sodium was sitting in his rocking chair on the front porch waiting for them. Sodium said, “I’m happy you made it so soon. I have a surprise for you on the other side of my house.” When they turned the corner of Sodium’s lovely home, they saw what Sodium had ready for them. It was a large bubble, different from the one Guy had used to fly around Periodic Table Land. They walked closer to get a better view of it.

As Sodium and Professor Terry began discussing details about the bubble, Guy wandered over to look more closely at the new flowers Sodium had planted in his rock garden. He spotted a frog swimming in the little pond in the middle of the garden. While Guy was enjoying nature, Professor Terry climbed into the bubble to check out the interior and the controls in the bubble. When Guy returned from admiring the garden, Professor Terry seemed to have vanished. Guy said, “So where’s Professor Terry?”

Sodium said, “Professor Terry has not gone anywhere, she is in the magic bubble.” Guy was looking right at the clear bubble, yet he couldn’t see any sign of her.

Sodium said, “Stick your head into the bubble, Guy.” He looked in, and there was Professor Terry big as life sitting inside. Sodium explained, “This magic bubble will make you and Professor Terry invisible.” Guy’s face lit up as his imagination ran wild with possibilities. “Professor Terry wants you to learn how compounds are formed. This magic bubble will allow you to be really close to where the atoms are becoming compounds without being in the way. You can be right up close and you will not interfere with what is happening. There is even a button on the control panel which will make the entire bubble invisible. Now, climb into the magic bubble and fly to 7th Street. I’ll be there soon.”

Professor Terry waved the magic wand, the bubble rose straight up, and then floated effortlessly over the trees, moving quickly and quietly toward 7th Street. There they would hopefully witness Sodium and Chlorine create their first compound. The atoms have been sad for so long that seeing them form a compound and become Happy Atoms will be an historic event in Periodic Table Land. “Let’s set up where the action will take place,” said Professor Terry.

Sodium Visits Chlorine

Not too much later, Sodium arrived in his bubble. His family’s light green balloons lowered him to a gentle landing on the grassy field behind 7th Street. He walked the short distance to 7th Street’s entrance on the trail through the pine forest.

Chlorine was busy planting colorful flowers at the base of the Halogen’s flag pole. The lovely Halogen flag was hanging limply above as the wind had died down.

Sodium cheerfully greeted Chlorine. “Hi there. Do you remember me? I’m that element who stopped by to see you while I was searching for a Happy Atom.” “Ah yes, and looking at your smiling face, I would guess that you found one.” “Well, yes. I found more than one. As a matter of fact, I found a whole family of Happy Atoms on 8th Street. Astatine was right when he said he heard a faint sound of laughter coming from 8th Street.” He then recounted the whole exciting story of his discovery. “Now I’m here to share with you a plan I’ve devised to make us all Happy Atoms like the ones I found on 8th Street. As you recall, I had just about given up hope of ever finding a Happy Atom. Then as I landed behind the forest at the top of 8th Street I found what I had been looking for—Happy Atoms, a whole family that was happy. I stayed until I found out their secret. Now I’m here to let you know how to become a Happy Atom. I have a plan for you.”

Sodium Explains How to Become a Happy Atom

"Let me hear your plan. We have been sad too long," said Chlorine.

"OK," said Sodium. "Becoming a Happy Atom is all about your Outside Energy Level. If that level is complete, an atom can be happy. It's that simple."

Chlorine said, "How does an Outside Energy Level get complete?

Sodium explained, "If we have only one energy level, the level is complete with only 2 electrons. If an atom has more than one energy level, then it is complete with eight electrons. Count how many electrons you have in your Outside Energy Level."

Chlorine said, "We live on 7th Street, Group 7A on the Periodic Table. So all the Halogens have seven electrons in our Outside Energy Level."

Sodium said, "That's why you are sad. To be happy you need 8 electrons. You are missing one electron. You have an empty space in your Outside Energy Level. You need to get one more electron to be complete."

Chlorine looked at his Outside Energy Level and said,"You are right. if we need 8 electrons to be complete, all the elements here on 7th Street need to get one more electron. Then, we will become a Happy Atom."

"It's even better than that Chlorine. Together we are going to create something new, a compound. It's very exciting."

Chlorine glanced again at his outside energy level and saw that one empty space that needed to be filled if he was going to be a Happy Atom and form a compound. A ray of hope began to grow in this sad little atom.

Sodium said, "You catch on fast. I have that one electron that you need. Look at the arrow pointing to that one electron I have in my Outside Energy Level. This is the electron I will give you to make your Outside Energy Level complete. Notice what I'm left with when I give away this electron."

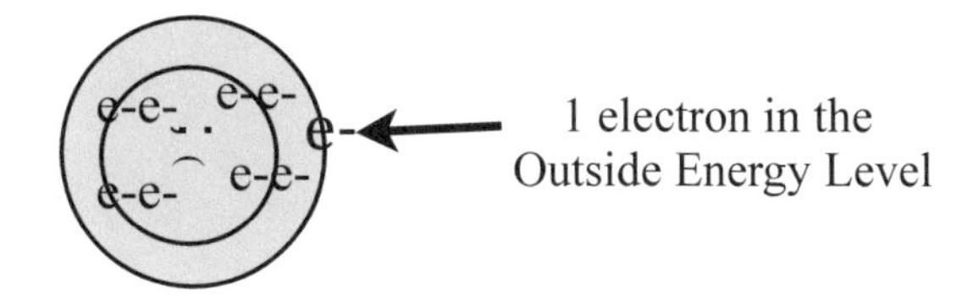

Chlorine looked closely, and said, "I see you have a complete energy level underneath that energy level with the one electron that you want to give to me. Your hidden energy level has 8 electrons in it. I counted them. It's complete."

"That's right," said Sodium. "My one electron will make us both Happy Atoms. I will be a Happy Atom when I give away that one electron, because that hidden complete energy level will pop up and become my Outside Energy Level. You will have 8 electrons when you get my one electron, and your Outside Energy Level will be complete. Then both of us will become Happy Atoms."

Happy Atoms, Compounds and Charges

Professor Terry turned to Guy and said, "There's even more to becoming a Happy Atom and forming a compound than that. Do you ever play with magnets?"

Guy said, "Of course, I love playing with magnets. It's so much fun. My magnets pull together, and when they get close enough I hear a click and they stay together. I can feel the force that pulls them together."

"Well, Guy" said Professor Terry, "when Sodium and Chlorine become Happy Atoms they're like your magnets. They pull together and click and that's how they stay together and become a new compound."

"Think about this. The ends of your magnets are marked with a + and a –sign."

Guy said, "I know. My father told me that one end is the positive + end and the other end is the negative – end. It's the + and – ends of the two magnets that pull together. My father said that they have opposite charges. The + and – ends of the magnet have some kind of an attractive force making them pull together."

"This is something like the force that Sodium and Chlorine will have when they get complete Outside Energy Levels. Open your Book 2, and take out your Bohr models of Sodium and Chlorine. I'll show you how they get opposite charges. Sodium gets a positive + charge and Chlorine gets a negative – charge. That's how they acquire a force like your magnets that holds them together in a compound. When they pull together, that's when they become Happy Atoms. The way they are right now, they are sad little atoms because they have no fun charges. Atoms have the same number of positive + protons as they have negative – electrons. You learned that your first day visiting the elements. All the elements have the same number of positive + protons and negative – electrons as their Atomic Numbers."

Atoms Have No Charge

Sodium's Atom

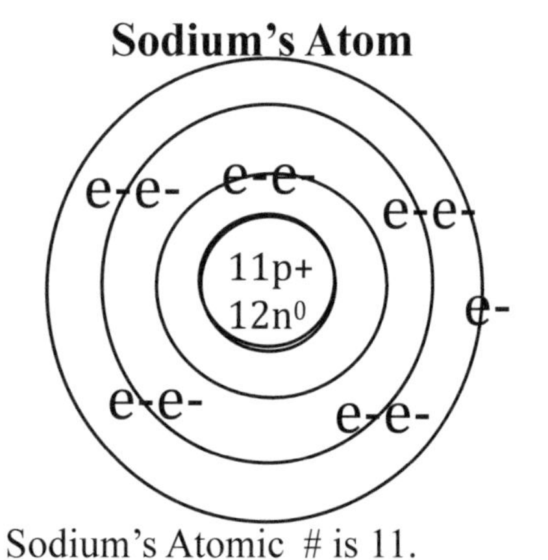

Sodium's Atomic # is 11.
So he has 11p+ 11e–

Chlorine's Atom

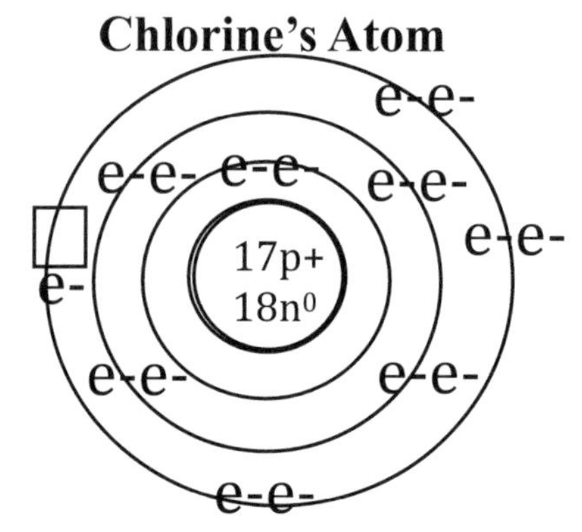

Chlorine's Atomic # is 17.
So he has. 17 p+ 17 e

"Guy, these are the atoms of Sodium and Chlorine. These atoms have no charge because the + charges equal the – charges. That means no charge. No fun. They can't attract each other like your magnets." She paused for a moment and then continued.

How Atoms Get a Charge

"Look how Sodium Na gets a charge when he gives away his 1 Outside Energy Level electron to Chlorine. Count the number of + protons and - electrons each has now."

Sodium Na^{+1}

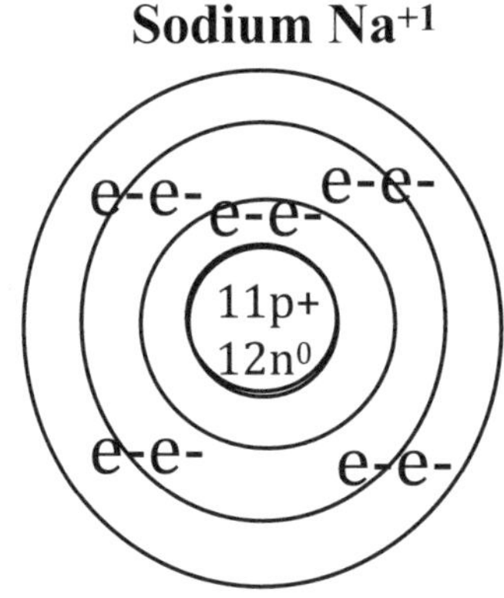

Chlorine Cl^{-1}

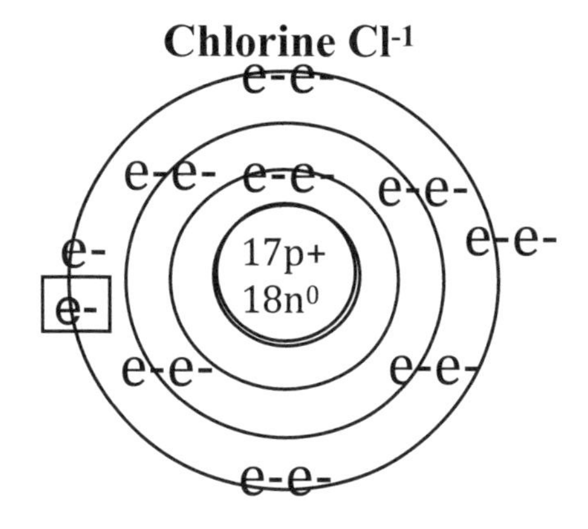

Sodium now has 11 p+ and 10 e–
Sodium has 1 more positive proton p+
So his charge is positive 1 or 1 p+

Chlorine now has 17 p+ and 18 e–
Chlorine has 1 more negative electron e–
So his charge is negative 1 or 1 e–

Once an atom gives away the Outside Energy Level electron that keeps him from having a complete Outside energy level, he ends up with more protons p+ than electrons. He gets a positive charge. This is similar to the force you feel in your magnet. Once the other atom that's missing an electron to be complete, takes in that electron, he gets a negative charge.

positive Na^{+1} **+** **–** Cl^{-1} **negative**

$Na^{+1}Cl^{-1}$

"Remember how the force of the magnets pulling together felt? Well when atoms get opposite charges, they pull together and a compound is formed. When this happens we say the atoms become Happy Atoms. Charges make them happy. It's the pull of the opposite charges that brings the atoms together to form a compound. That's the rest of the story Guy. Now you know it's all about the positive protons attracting the negative electrons that causes the compound to stay together."

"I'll add one more idea here. When an atom gets a charge, it is called an Ion. When Ions form a compound it's called Ionic bonding. This kind of compound formation happens when the atoms give away or get electrons. The compound is held together by the force of these opposite charges (+ and –) attracting each other. Most of the compounds you will be learning about are formed this way."

"When we study the Boron Family I'll explain a little about one other way compounds are formed called Covalent Bonding. Now let's get back to watching what Sodium and Chlorine are doing."

Sodium continuing to explain to Chlorine how they could become Happy Atoms said, "Now the challenge is deciding what we must do to give you my electron. I have a plan. Of course, it is a hypothesis. It needs to be tried to see if, indeed, it will work. Let's see if it really can be done. Are you willing to try?"

Chlorine, impatient as he was to end the sadness that had engulfed his whole family, agreed to begin. "What are we waiting for? I want to be a Happy Atom. Let's get started."

Nearby, Professor Terry and Guy, invisible in their bubble, looked on in anticipation. As Chlorine said this, Professor Terry whispered to Guy. "You are about to observe a momentous event. Becoming a Happy Atom is really the making of a new compound. So pay attention. You are going to see Sodium and Chlorine change, and become something new and different. They will not be sad little atoms any more. They will be Happy Atoms hidden away in that new compound that they form. After they're finished, you will have completed your first lesson in compound formation. Let's watch."

Sodium and Chlorine Become Happy Atoms

Sodium had mapped out the whole plan on paper and placed it on Chlorine's table. Chlorine bent over the table and carefully read the page of directions several times. Finally, he indicated that he was ready to begin saying, "OK, I'm ready but I have one question. It talks about the metal and non-metal lining up in special places. How do I know if I'm a metal or a non-metal?"

Sodium said, "Let me tell you the difference between a metal and a non-metal.

"A **metal is an atom that has extra electrons to give away or share in a chemical reaction**. That makes me a metal."

"Non-metals are atoms that are missing electrons and they need to get electrons in a chemical reaction. That makes you a non-metal."

So Sodium is the metal with an electron to give away. Chlorine is the non-metal needing to get an electron.

Watch Sodium and Chlorine Become Happy Atoms

The **metal** always stands on the **left**

The **non-metal** always stands on the **right**.

Sodium(Na)
the metal
stood on the **left**

Chlor***ine*** (Cl)
the non-metal
stood on the **right**

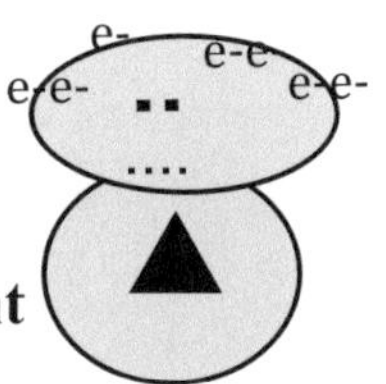

Then they begin to move closer and closer and closer.

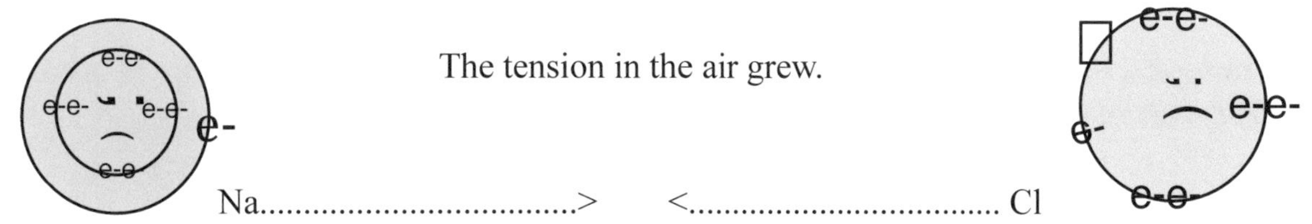

The tension in the air grew.

Suddenly, when they were really close, Sodium's 1 electron jumped into the empty space in Chlorine's Outside Energy Level. As this happened there was a breathtaking sound. It was difficult to describe. The sound was a combination of the explosion of fireworks, cymbals clashing, and the fanfare of a dozen trumpets blaring a welcoming tribute. This sound occurred precisely when Sodium's electron jumped into the empty space in Chlorine's Outside Energy Level.

It was magical. Sodium and Chlorine were changed the moment Sodium's electron popped into Chlorine's Outside Energy Level. Their atoms had joined together and they were now a *compound.* The two little elements, Sodium and Chlorine tucked away inside the compound became happy. Look at their smiling happy faces!

They are now, Happy Atoms.

There's more, they are now a compound!

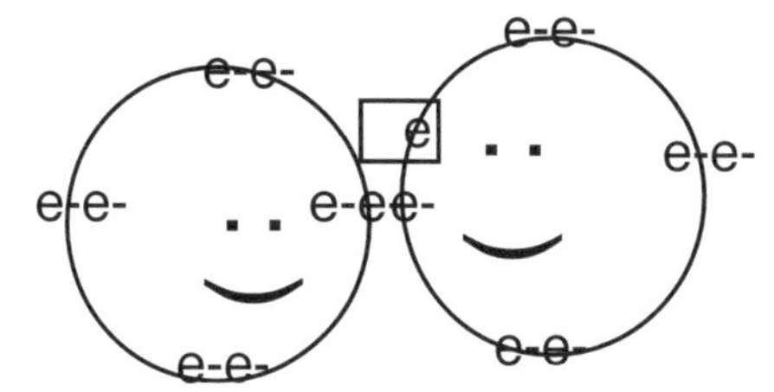

They instantly get a **new name**: ....Sodium Chlor***ide.***
Their symbols are now written together, as the compound's **formula...**NaCl

Notice the element on the right, the non-metal, changed the ending of his name.
*ine changed **to ide**—* Chlor***ine*** became Chlor***ide***

Here's how we write what happened in words and diagrams:

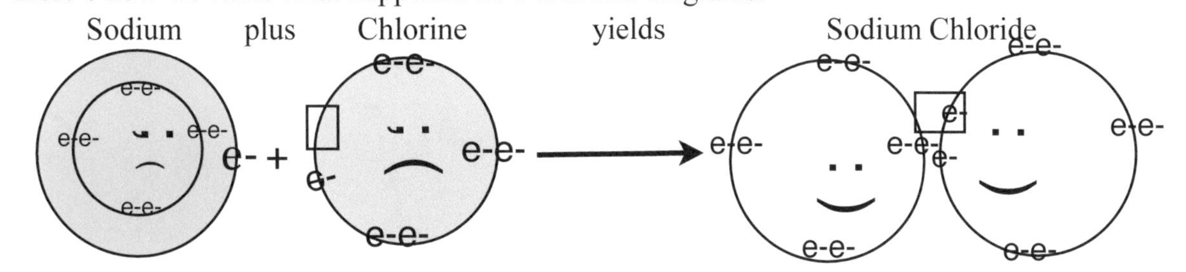

So delighted Chlorine exclaimed, "This is so exciting! Our experiment worked. Now, both our families will be able to become Happy Atoms." Sodium and Chlorine discussed how to get the Alkali Metal and Halogen families together to make compounds and become Happy Atoms.

When the plan was complete, Sodium jumped into the family bubble with those beautiful light green balloons. He wished it could go faster, but he had to be satisfied with effortlessly gliding through the air. It was a lot faster than walking. This is a chance to learn the patience that Guy was telling me about. So, he decided to relax and enjoy the ride back to 1st Street. He thought about how happy his family would be, when he got home and told them his plan worked. Now, all the Alkali Metals will be able to become Happy Atoms. So will the Halogens.

Guy Learns the Basics of Compound Formation

Professor Terry then told Guy, "I want you to notice how they wrote in words and diagrams what happened forming the compound. Eventually you will be able to write this statement using chemical symbols and formulas. It will be called a Chemical Equation."

"This was exciting!" Guy hidden away in his invisible bubble had watched something new and wonderful being formed—a compound. He turned to Professor Terry and said, "I remember wondering how there could be so much variety in this world when the whole universe is made up of only 92 elements. I guess forming compounds is one way the 92 elements can create new and different matter in our world." Guy now knew a little bit more about his magnificent world.

Professor Terry said, "Guy, this is just the beginning. There is much more to learn to really understand how compounds are formed." Guy wondered how much more.

"Guy, I would like you to take out the Periodic Table displaying metals and non-metals."

THE PERIODIC TABLE OF THE ELEMENTS

PERIODS ↓ GROUPS →

FORMS OF MATTER
Metals - red
Metalloids - Orange
Non-metals - Yellow

Periods	1A	2A	3B	4B	5B	6B	7B	8B	8B	8B	1B	2B	3A	4A	5A	6A	7A	8A
1.	1 H 1.00																	2 He 4.00
2.	3 Li 6.94	4 Be 4.01											5 B 10.8	6 C 12.0	7 N 14.0	8 O 16.9	9 F 18.9	10 Ne 20.2
3.	11 Na 22.9	12 Mg 24.3											13 Al 26.9	14 Si 28.1	15 P 30.9	16 S 32.1	17 Cl 35.5	18 Ar 39.9
4.	19 K 39.1	20 Ca 40.0	21 Sc 44.9	22 Ti 47.9	23 V 50.9	24 Cr 51.9	25 Mn 54.9	26 Fe 56.8	27 Co 58.9	28 Ni 58.7	29 Cu 63.5	30 Zn 65.4	31 Ga 69.2	32 Ge 72.6	33 As 74.9	34 Se 78.9	35 Br 79.9	36 Kr 83.8
5.	37 Rb 85.5	38 Sr 87.6	39 Y 88.9	40 Zr 91.2	41 Nb 92.9	42 Mo 95.9	43 Tc {98}	44 Ru 101.	45 Rh 102.	46 Pd 106.	47 Ag 107.	48 Cd 112	49 In 114	50 Sn 118	51 Sb. 76	52 Te 126.	53 I 126.	54 Xe 131.
6.	55 Cs 133	56 Ba 137	57 - 71	72 Hf 178	73 Ta 181	74 W 184	75 Re 186	76 Os 190	77 Ir 192	78 Pt 195	79 Au 197	80 Hg 200	81 Tl 204	82 Pb 207	83 Bi 208	84 Po 209	85 At 210	86 Rn 222
7.	87 Fr 223	88 Ra 226	89-103	104 Rf 267	105 Db 268	106 Sg 271	107 Bh 272	108 Hs 270	109 Mt 278	110 Gs 281	111 Rg 280	112 Cn 285	113 Nh 284	114 Fl 289	115 Mc 288	116 Lv 203	117 Ts 294	118 Og 294

57 La 139	58 Ce 140	59 Pr 141	60 Nd 144	61 Pm 145	62 Sm 150	63 Eu 152	64 Gd 157	65 Tb 159	66 Dy 163	67 Ho 165	68 Er 167	69 Tm 169	70 Yb 173	71 Lu 175
89 Ac 227	90 Th 232	91 Pa 231	92 U 238	93 Np 237	94 Pu 234	95 Am 243	96 Cm 247	97 Bk 247	98 Cf 251	99 Es {52	100 Fm 257	101 Md 258	102 No 259	103 Lr 262

Metals and Non-metals

While Professor Terry and Guy waited for the Alkali Metals to return, Professor Terry took this time to teach Guy the rules involved in creating compounds.

Guy held up the chart showing metals and non-metals, saying, "Here it is! On the chart the metals are colored red and the non-metals are colored yellow."

She reminded Guy, "When I first gave you that chart, I told you that I would explain how scientists decide whether an element is a metal or a non-metal when it was important for you to know this. Now is the time for you to learn the real meaning of metals and non-metals. It is absolutely necessary for you to know this, if you are to understand how elements combine to form compounds and how you will write the names and formulas for the new compounds."

She continued, "You know that the elements colored red on the Periodic Table are the Metals. Now I will add more to that. **Metals have electrons to give away to achieve a complete Outside Energy Level**. You know that the elements, colored yellow on that chart are the non-metals. I will now add more to that. **Non-metals need to take in electrons to have complete Outside Energy Levels.** This is rather simple to understand, but important to remember because when you are going to write the name and formula for the compound you must know which element is the metal and which is the non-metal. It's the **metal** that is written **on the left** in both the compound's name and the formula. The **non-metal** is written **on the right** in the compound's name and formula**.** You also must remember that **the non-metal changes the ending of his name to *__ide__*."**

Sodium Combining with Chlorine Reviewed

Professor Terry continued. "Let's review what happened to Sodium and Chlorine using the above facts about metals and non-metals."

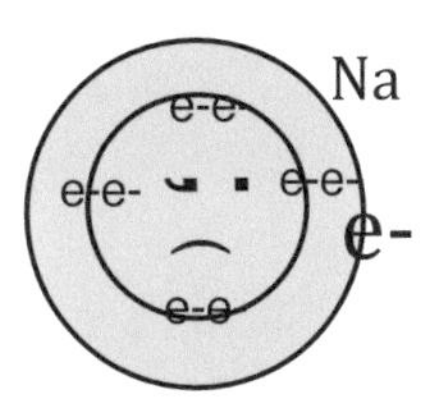

"Sodium, the metal, is the element giving away electrons. The metal is always written on the left in the formula and in the name of the compound. So, the compound's name was **Sodium** Chloride; the formula **Na**Cl. Sodium and Na were written on the left. Most of the Metals are found in Groups 1A, 2A, 3A, and in the B Groups. Look at the red portion of the table to see the elements that are metals. They all follow these rules for writing the name and formula for their compounds"

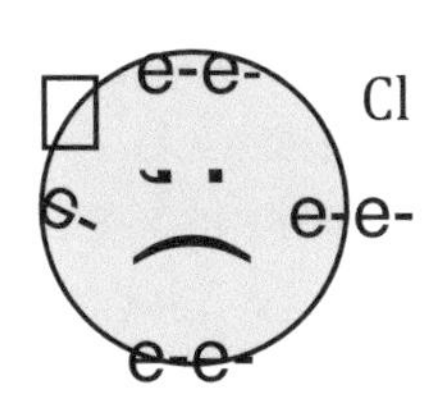

"Chlorine, the non-metal, is the element that is missing electrons and needs to get electrons when forming a compound. The non-metal, is always written on the right in the formula and in the name of the compound. Chlorine**,** the non-metal is written on the right in formula: Na**Cl;** on the right in the compound formed: Sodium **Chloride.** Also notice Chlor**ine** changed his name to Chlor**ide.** The non-metal changes its ending from **ine** to **ide**,. Most non metals are in Groups 5A, 6A and 7A and all follow these rules."

Guy said, "Let me see if I understood what you said."

Sodium (Na), the Metal in Group 1A, is the element that has the electron to give away**.** He starts on the left. Sodium's electron jumps into Chlorine's Outside Energy Level. **Sodium (Na), the metal** is written on the left in the formula, **Na**Cl for the compound, and on the left in the name of the compound, **Sodium** Chloride.

Chlorine (Cl), the non-metal in Group 7A, needs to get an electron. He starts on the right. He moves toward Sodium and takes in Sodium's electron. **Chlorine (Cl)** is written on the right in the formula, Na**Cl** and also written on the right in the name of the compound, Sodium **Chlor*ide*.** Na Cl is the formula for the compound**,** Sodium Chloride. I see that Chlorine, the non-metal, written on the right, changed the ending of his name from *ine* to *ide,* Chlor**ine** becomes Chlor**ide**. It's always the element on the right, the non-metal, that changes the ending of his name.

"You got it, Guy. Follow these instructions, and you will be able to show how a compound is formed by joining any 2 elements. When these elements give away and get electrons in forming compounds, they get a charge."

Guy said, "I remember you taught me all about the atoms becoming ions when they get a charge. Then they attract each other like the ends of a magnet. The opposite charges are what create the force that hold the compound together. Atoms with a charge are called ions. So, when they form compounds this way, it's called ionic bonding."

"Good job, Guy. I didn't really expect you to understand Ionic Bonding but you do. The atoms get a charge and it's their attraction like a magnet that holds the compound together. Atoms with a charge are ions. That's why they call it Ionic Bonding."

While they waited for the Alkali Metals to show up; and as they were taking a long time, Professor Terry had time to teach Guy something else. "Do you remember how Chlor***ine*** changed his name to Chlor***ide*** when Sodium and Chlorine formed a compound? Well the rest of the Halogens are going to change the ending of their names in the same way when they form compounds."

Fluor***ine***	will become	Fluor***ide,***
Brom***ine***	will become	Brom***ide***,
Iod***ine***	will become	Iod***ide***
Astat***ine***	will become	Astat***ide***

"That's a lot for you to remember; but as I said before, when you watch compounds being formed over and over, it will become second nature to you."

Guy said, "I've really learned a lot. Looking at that red/yellow Periodic Table showing metals and non-metals, and watching Sodium and Chlorine become Happy Atoms has helped me understand why elements combine to form compounds." Guy said,"I've learned so much already. I know so much more about my world. When I sit in my favorite spot on the mountain behind my parent's cabin watching the night sky, I have so much more to think about. The stars will remind me of the elements Hydrogen and Helium—the elements that make up so much of the sun and other stars. I will imagine their Bohr models with the electrons whirling around the nucleus like the planets out there in space. I will realize the cool night air I'm breathing contains both Oxygen and

Nitrogen atoms. Then I will look at the trees and bushes around me and imagine the Nitrogen and Phosphorus atoms doing their work in the soil making these trees and bushes on the mountain grow stronger. I've only begun to learn chemistry, and already I know so much more about my world."

Good News For Sodium's Family, the Alkali Metals

While Professor Terry continued her lesson on compounds with Guy, Sodium was on his way back to 1st Street. He couldn't wait to get his good news to his Alkali Metal Family, and get them ready to come back to 7th Street. Imagine my whole family will soon learn to become Happy Atoms. Sodium was so excited.

Sodium landed his bubble, jumped out and ran as fast as he could to find his family. They were all gathered together in Lithium's rose garden. Sodium quickly announced, "My experiment worked! Chlorine and I became Happy Atoms. I was able to give away my outside electron to Chlorine and when I did we both became Happy Atoms. It was so wonderful! Now I know it's possible for all of you to become happy like those elements on 8th Street. We don't have to be sad anymore. Each one of you has that one electron needed by the Halogens to become a Happy Atom. There is every reason to believe that our whole family, and all the Halogens will soon be happy like the elements on 8th Street. None of us will be sad any more."

All the Alkali Metals cried out. "Oh Sodium! We love you. You are so wonderful. When can we go? We have been sad too long. It's time to be happy."

Back on $7^{th.}$ Street Professor Terry and Guy waited for Sodium to return with the whole Alkali Metal Family.

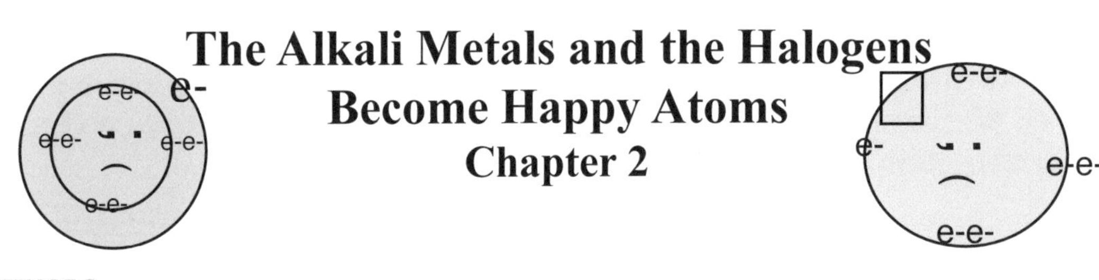

The Alkali Metals and the Halogens Become Happy Atoms

Chapter 2

THE PERIODIC TABLE OF THE ELEMENTS

Periods ↓ / Groups →

Periods	1A	2A	3B	4B	5B	6B	7B	8B	8B	8B	1B	2B	3A	4A	5A	6A	7A	8A
1.	1 **H** 1.00																	2 **He** 4.00
2.	3 **Li** 6.94	4 **Be** 4.01											5 **B** 10.8	6 **C** 12.0	7 **N** 14.0	8 **O** 16.9	9 **F** 18.9	10 **Ne** 20.1
3.	11 **Na** 22.9	12 **Mg** 24.3											13 **Al** 26.9	14 **Si** 28.0	15 **P** 30.9	16 **S** 32.0	17 **Cl** 35.5	18 **Ar** 39.9
4.	19 **K** 39.1	20 Ca 40.0	21 **Sc** 44.0	22 **Ti** 47.9	23 **V** 50.9	24 **Cr** 51.9	25 **Mn** 54.9	26 **Fe** **56**	27 **Co** 58.9	28 **Ni** 58.6	29 **Cu** 63.5	30 **Zn** 65.3	31 **Ga** 69.2	32 **Ge** 72.6	33 **As** 74.9	34 **Se** 78.9	35 **Br** 79.9	36 **Kr** 83.7
5.	37 **Rb** 85.5	38 **Sr** 87.6	39 **Y** 88.9	40 **Zr** 91.2	41 **Nb** 92.9	42 **Mo** 95.9	43 **Tc** {98}	44 **Ru** 101	45 **Rh** 103	46 **Pd** 106	47 **Ag** 107	48 **Cd** 112	49 **In** 114	50 **Sn** 119	51 **Sb** 122	52 **Te** 126	53 **I** 126	54 **Xe** 131
6.	55 **Cs** 132	56 **Ba** 137.	57 - 71	72 **Hf** 178	73 **Ta** 180	74 **W** 183	75 **Re** 186	76 **Os** 190	77 **Ir** 192	78 **Pt** 195	79 **Au** 196	80 **Hg** 200	81 **Tl** 204	82 **Pb** 207	83 **Bi** 208	84 **Po** 209	85 **At** 210	86 **Rn** 222
7.	87 **Fr** 223	88 **Ra** 226	89-1 03	104 Rf 267	105 **Db** 268	106 **Sg** 271	107 **Bh** 272	108 **Hs** 270	109 **Mt** 278	110 **Gs** 281	111 **Rg** 280	112 **Cn** 285	113 Nh 284	114 **Fl** 289	115 **Mc** 288	116 **Lv** 203	117 Ts 294	118 Og 294

The Alkali Metals in Group 1A were excited! When they calmed down, Sodium said, "We can leave as soon as you are ready." They all screamed with excitement; they couldn't wait to become Happy Atoms. Sodium continued, "Calm down now. I'll give you ten minutes to get ready, and then let's gather at the head of our street beside the tarmac. I'm going there now to ready our balloons for the flight."

With a wave of his magic wand, the bubble below the family balloons became large enough to carry all the Alkali Metals to 7th Street, the home of the Halogens, Group 7A on the Periodic Table. Sodium added more of the light green balloons to the family bus to help it fly high into the blue sky. The bubble beneath had become gigantic. All the elements slid into the bubble, and there was happy chatter. Sodium waved his magic wand and pointed it in the direction of 7th Street. They were off to become Happy Atoms.

They all looked down at the unique houses below. It was a pleasant trip. They passed B Avenue, and 3rd Street. Before they knew it, they had reached 7th Street. The family bus landed safely in a clearing in the woods behind 7th Street. Out poured Lithium, Sodium, Potassium, Rubidium, Cesium and Radium. They all ran to the edge of the pine forest. Sodium set out on foot to meet Chlorine and arrange for his family to become what they had hoped to be for so long—Happy Atoms.

The Alkali Metals remained at the edge of the woods. They were all full of anticipation as they waited for Sodium to get things organized.

Before he left, Sodium had said, "I'll hold both my hands over my head as a signal, when it's time to come."

Sodium greeted Chlorine, "I'm back, and I brought my entire family with me. They are ready to become Happy Atoms."

Chlorine said, "I told my family all about our successful experience, and they can't wait to see if they can become Happy Atoms, too. I'll go get all my Halogens."

The Alkali Metals saw Sodium's signal, and they came running. They all gathered together waiting for the Halogens. They were so excited about what was going to happen next.

Chlorine returned with Fluorine, Bromine, Iodine and Astatine. They were chatting with each other, and their nervous energy showed that they couldn't wait to begin.

Both the Alkali Metals and the Halogens were eager to become Happy Atoms, but they were scared too. "Who will be the brave ones to go first?" said Sodium. Finally Potassium came forward for the Alkali Metals.

The Halogens, Fluorine and Bromine, were trying to push each other forward saying, "You go, Bromine."

"No, you go," returned Fluorine.

Iodine, annoyed that they were acting so silly, called out, "I'll go first." Then, he made his way through the group to the front.

Potassium and Iodine Become Happy Atoms

Professor Terry and Guy were nearby in their special bubble cloaked with invisibility. The professor turned to Guy and said, "It's time for a quick review. Remember: A **Metal is an element that has an electron to give away when forming a compound.** That makes all the Alkali Metals in Group 1A, metals. They have one electron to give away."

A **non-metal is an element that is missing an electron and needs to take in electrons when forming a compound.** That makes all the elements in Group 7A, non-metals. They need to take in one electron to have a complete Outside Energy Level."

"Let's watch how these Alkali Metals combine to form compounds with the Halogens. We're about to see the metal, Potassium, combine with the non-metal, Iodine, to form a compound. I brought my recorder along, and I'm going to narrate the entire action as we observe it happening. I want you to pay close attention. We can listen to the recording later, and it will mean so much more to you because you saw it happening with your own eyes. The action is about to start. Pay close attention, Guy."

Sodium and Potassium moved out into the center of the grassy area. Guy and Professor Terry were in their invisible bubble as close as possible. Sodium reminded Potassium and Iodine of the rules, and the action began.

The **metal,** the element with an **electron to give away** always stands **on the left**. That makes Potassium a metal because he has an electron to give away. So he stands on the left.

The **non-metal,** the element that needs to **take in an electron** to have a complete Outside Energy Level always stands **on the right**. That makes Iodine a non-metal, and he stands on the right.

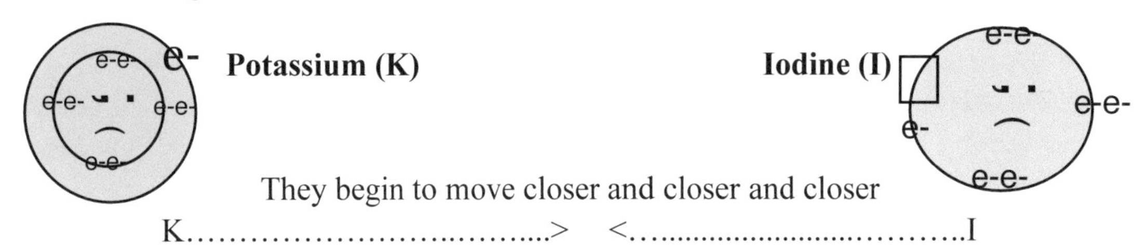

They begin to move closer and closer and closer

K…………………………..……...> <…...........................………..I

Suddenly, when they get really close, Potassium's electron jumps into the empty space in Iodine's Outside Energy Level. The moment this happens, there is a breathtaking sound. It was magical! As Potassium's electron moved into Iodine's Outside Energy Level, Potassium's hidden energy level, that was complete, popped up. Both atoms became COMPLETE. Look at their beautiful smiling faces! They became Happy Atoms.

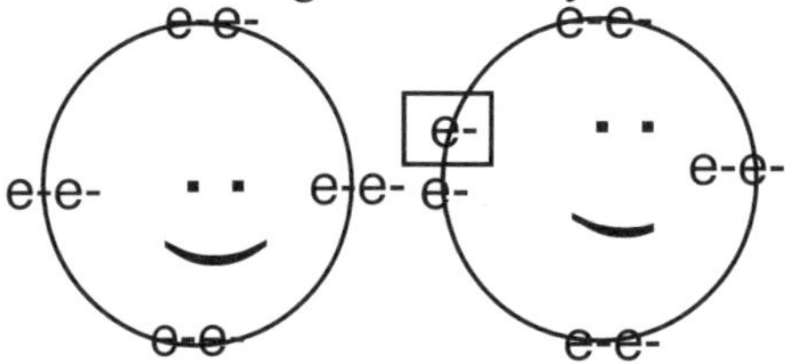

They were more than just happy. Now together the atoms were a new creation. They were no longer just Potassium and Iodine. They were now a COMPOUND with a new name and formula.

Their new name is...................Potassium Iod**ide.**
Their formula is………………………..KI.

The non-metal element was the one to change the ending of his name.
*ine changed to ide, Iod**<u>ine</u>** became Iod**<u>ide</u>***

Here's how to write what happened in words and in diagrams.

Potassium plus Iod**ine** yields Potassium Iod**ide**

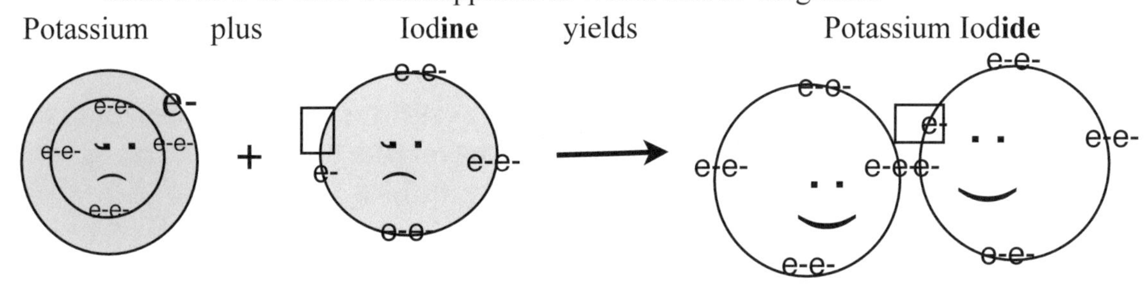

It took just one atom of Potassium and one atom of Iodine to become a compound, a Happy Atom. That's why we need only one symbol for each element in the formula, and we write these symbols together. We write the formula as KI because it was formed when one atom of Potassium, K joined with one atom of Iodine, I.

Guy turned to Professor Terry and asked, "Why don't we write a one beneath each element in the formula like this K_1I_1 to show there is only one atom of each element in the compound?"

Professor Terry explained, "We don't need to write one as a subscript. It's just not necessary. Since the symbol is there it is understood that it represents 1 atom of that element."

Rubidium and Bromine Form Happy Atoms

Professor Terry and Guy watched from the privacy of their bubble. "Let's watch another compound being formed. It will be done in the exact same way."

Sodium said, "Which two elements will go next?"

The Alkali Metal, **Rubidium** said, "I will."

The Halogen, Brom**ine** said, "I will." They lined up. Since they watched Potassium and Iodine, they felt confident they would be able to become Happy Atoms.

Sodium reminded **Rubidium and Bromine** of the rules.

The **meta**l has an electron to give away and stands on the left. That makes Rubidium from Group IA a **metal**.

The **non-metal,** missing an electron, needs to get an electron and stands on the right. That makes Bromine from Group 7A a **non-meta**l.

They begin to move closer and closer and closer.

Rb..............................> <........................Br

Suddenly, when they get really close, Rubidium's electron jumps into the empty space in Bromine's Outside Energy Level. Then Rubidium's hidden energy level that was complete, popped up. Both atoms became COMPLETE. Look at their smiling faces.

Now they are Happy Atoms.

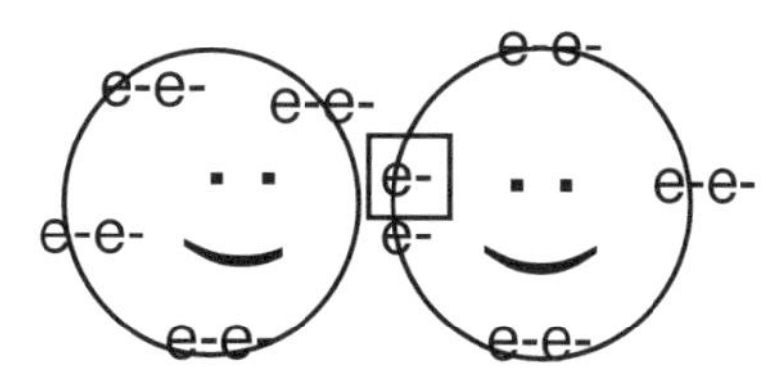

They were more than just happy. Now together, these Happy Atoms were a beautiful new creation. They had become a COMPOUND with a new name and formula.

Their new name is............................Rubidium Brom***ide***

Their formula is..................................RbBr

The non-metal element was the one to change the ending of his name.

*ine changed to ide. Brom**ine** became Brom**ide***

Here's how to write what happened in words and diagram.

Rubidium plus Bromine yields Rubidium Bro**mide**

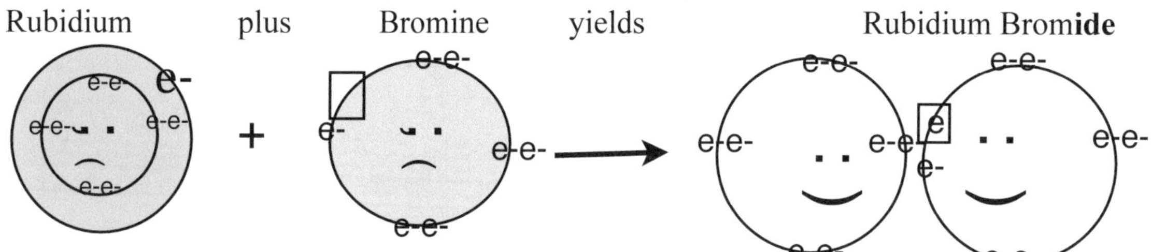

That's why we need only one symbol for each element in the formula, and we write these symbols together. We write formula as **RbBr** because this compound was formed when one atom of Rubidium joined with just one atom of Bromine.

Cesium and Fluorine Become Guy's Challenge

Professor Terry turned to Guy and asked if he could prove that he knew how an Alkali Metal and a Halogen could form a compound and become Happy Atoms. Guy agreed to try. So he got out of the bubble as he told Professor Terry, "I'll start by choosing the Alkali Metal **Cesium** and the Halogen **Fluorine** to form a compound. Then he turned to **Cesium and Fluorine** and reminded them of the rules:

"A **metal** gives away electrons to form a compound. Cesium is in Group 1A and has 1 electron to give away to make a compound. That makes Cesium a metal." Just to show Professor Terry he understood what happens when a metal gives away an electron Guy added, "When Cesium gives away this electron, his hidden complete energy level will pop up and become the complete Outside Energy Level an element needs to be a Happy Atom."

"A **non-metal** is missing an electron and needs to get an electron to make a compound. Fluorine is in Group 7A with 7 electrons in his Outside Energy Level. Fluorine needs to get 1 more electron to have the 8 electrons needed to have a complete Outside Energy Level. That makes Fluorine a non-metal."

Guy continued, "The **metal** always stands on the left.

The **non-metal** always stands on the right."

Guy watched them move into place. Then he continues explaining what happens.

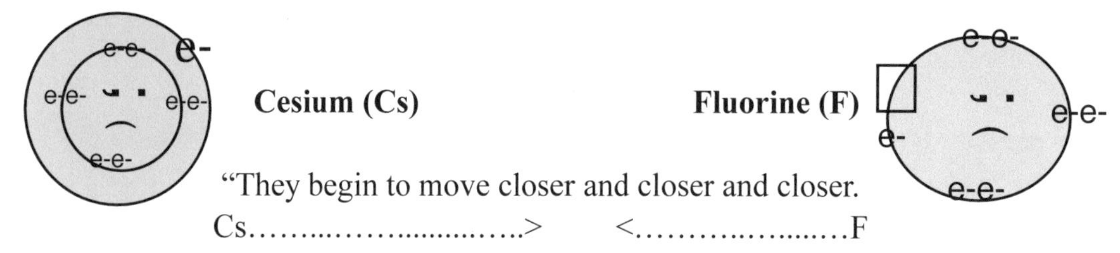

Cesium (Cs) **Fluorine (F)**

"They begin to move closer and closer and closer.

Cs……..……..........…..> <……….…….....…F

Suddenly when they get really close, Cesium's electron jumps into the empty space in Fluorine's Outside Energy Level. Fluorine then has 8 electrons and is complete. Cesium's hidden energy level that was complete pops up. They are now Happy Atoms."

Look at their smiling faces!

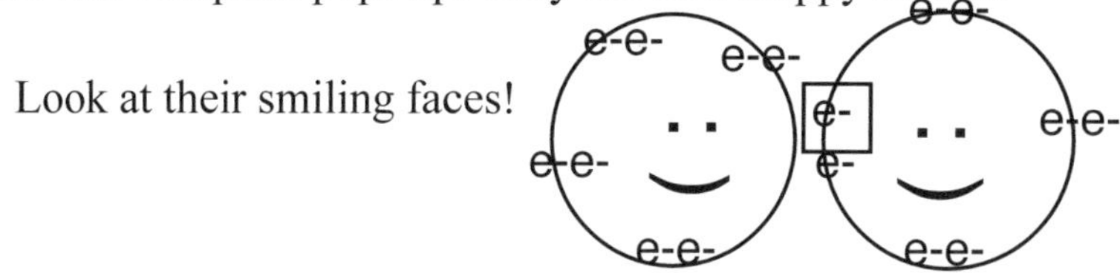

Guy continued, "They were more than just happy. Now together, they were a new creation. They were no longer just Cesium and Fluorine. They are a COMPOUND with a new name and formula."

Their new name is.....................................Cesium Fluor***ide***.

Their formula is...CsF.

The non-metal element changed the ending of his name, ine, to ide.

*Fluor**ine** changed the ending of his name and became Fluor**ide***

Here's how to write what happened in words and diagrams.

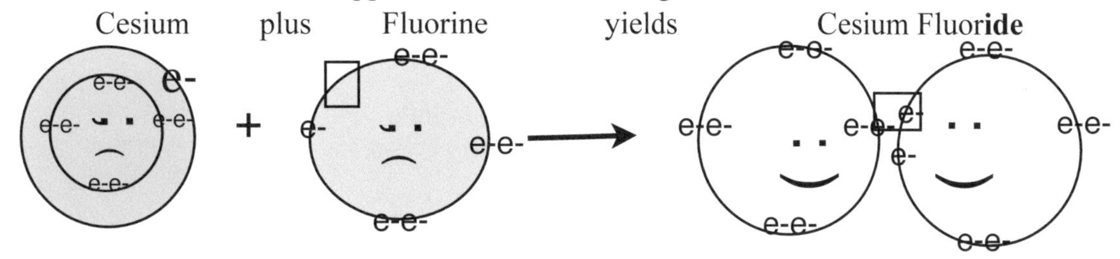

"It took only one atom of Cesium and one atom of Fluorine to become a compound—a Happy Atom. If we write the two symbols together as the formula, it means that it only took one atom of each to form the compound. So, **CsF** means one atom of Cesium joined with one atom of Fluorine to form the compound, Cesium Fluoride."

Observing that Guy definitely understood how the Cesium formed a compound with Fluorine, she felt he could understand that all of the Alkali Metals form compounds with the Halogens in the very same way. Guy was learning fast how compounds were formed.

All Alkali Metals Form Compounds and Become Happy Atoms

Professor Terry continued, "Guy, all the elements in Group 1A when forming compounds with the elements in Group 7A do it the same exact way. The Alkali Metals in Group 1A each has one electron to give away. Here are the Alkali Metals: Lithium, Sodium, Potassium, Rubidium, Cesium and Francium. The elements in Group 7A need to get just one electron. Here are the Halogen elements: Fluorine, Chlorine, Bromine, Iodine and Astatine. So, it just takes one atom of each element to form a compound. For the formula, we simply write their symbols together."

"I'll show you in words and diagrams how any Alkali Metal and Halogen join together to make a compound and become Happy Atoms."

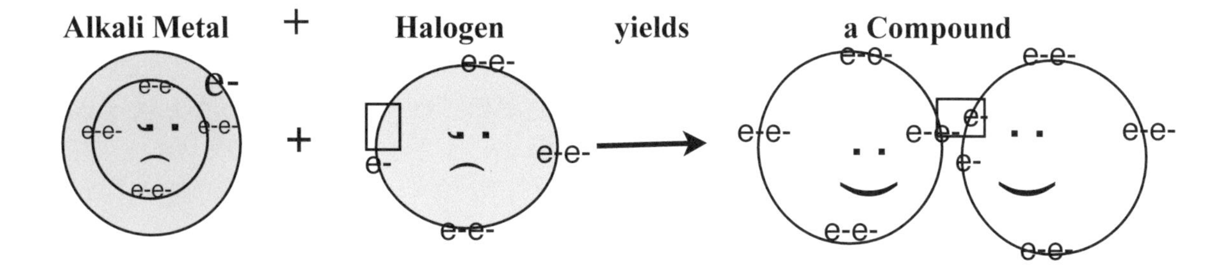

"Here are a few compounds made when one atom of an Alkali Metal joins with one atom of a Halogen. The whole process is the same when each element forms a compound."

Alkali Metal, Group 1A	**+**	**Halogen, Group 7A**	**⟶**	**Compound**	
Sodium (Na)	plus	Iodine (I)	yields.	NaI	Sodium Iod**ide**
Potassium (K)	plus	Fluorine (F)	yields	KF	Potassium Fluor**ide**
Cesium (Cs)	plus	Bromine (Br)	yields	CsBr	Cesium Brom**ide**
Rubidium (Rb)	plus	Astatine (At)	yields	RbAt	Rubidium Astat**ide**

Guy turned to Professor Terry and asked, "How many compounds can these two Chemical Families make?"

Professor Terry said, "Each of the six Alkali Metals

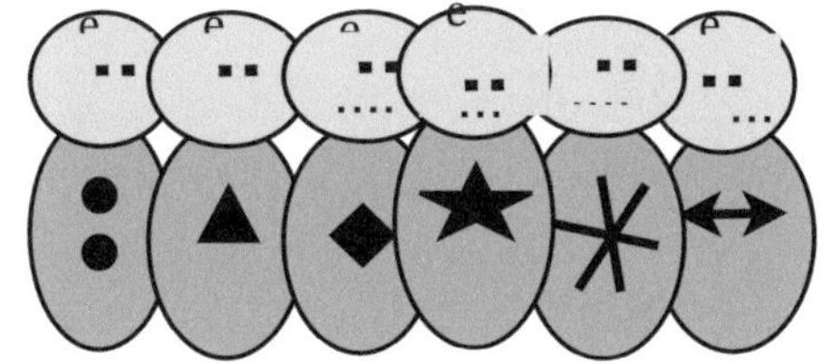

can join with each of the five Halogens

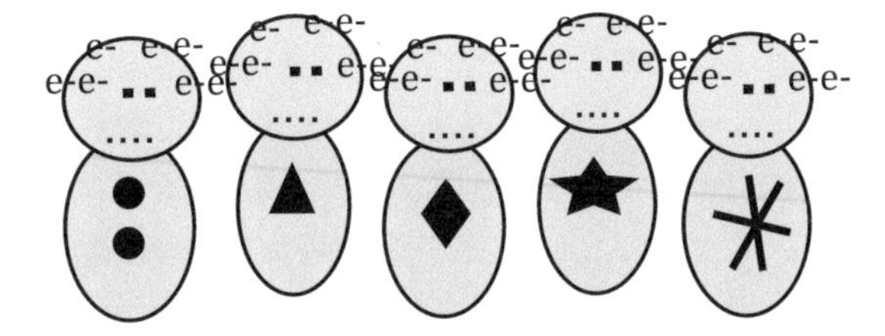

forming five different compounds each."

Sodium ever the teacher said, "I can make the 5 compounds Sodium Fluoride, Sodium Chloride, Sodium Bromide, Sodium Iodide and Sodium Astatide. Each of you can make 5 compounds: Fluorides, Chlorides, Bromides, Iodides and Astatides. Put the name of your element in front of these Halogens. That's the compound you'll make."

"If each of us makes 5 compounds we'll have 30 new compounds on 1st Street. Add it up. I make 5 compounds, Lithium makes 5 compounds, Potassium makes 5 compounds. You last 3 elements make 5 compounds. That's 30 compounds: 5+5+5+5+5+5 = 30. If you know math Sodium continued, 5 six times is 5 x 6 = 30."

"Wow! That's a lot of compounds," said Guy.

Professor Terry responded, "Yes, and that's a lot of happiness to add to Periodic Table Land. Now I have something to teach you about Lithium as we finish learning about how the Alkali Metals form compounds with the Halogens, Lithium is just a little different from the other members of this family. See if you notice what is different."

Lithium and Chlorine Becomes Happy Atoms

Professor Terry turned to Guy and said, "I saved Lithium to the very end because I wanted you to notice how very different Lithium is even though Lithium's final formula is just like the other elements in the Alkali Metal family."

Then Professor Terry asked if Guy knew any way that Lithium was different from the other members of Alkali Metal Family.

Before he could answer, Lithium interrupted. "I can tell you how I'm different. I'm the only Alkali Metal that has only one energy level, after I give away my electron, and that energy level has only 2 electrons in it. Take a look at my atom."

"I have only 2 electrons where all the other members of my family have had 8 electrons. I know you promised me that I will become a Happy Atom, but I'm still worried that I will not because I'm not like the rest of my family."

A tear dropped out of Lithium's eye.'

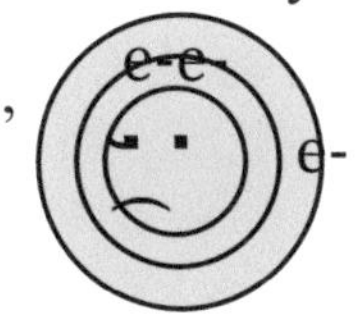

Professor Terry said, "Don't worry Lithium. Here's a Bohr model of your atom."

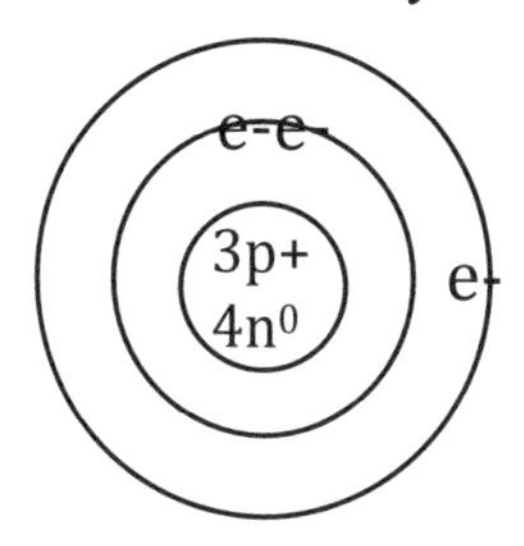

"Look at the energy level next to your nucleus. It is complete with only 2 electrons. That's why we stopped putting more electrons there and started a new energy level. It was complete. It could't hold any more electrons. All energy levels right next to the nucleus can only hold two electrons. It can not hold any more electrons. If there are any more electrons, they need to be placed in the next energy level. That means the energy level next to the nucleus is complete with 2 electrons and complete is all it takes to be a Happy Atom. I promise you'll be as happy as the rest of your family because it's not how many electrons that's important. It's that the Outside Energy Level needs to be complete and yours is complete with only 2 electrons. By the way, there are two other elements Beryllium and Boron in different families that are like you. They also will become Happy Atoms with only 2 electrons in their Outside Energy Level."

"Lithium stop worrying. I'll show you how you and Chlorine can become Happy Atoms." She began by reminding Lithium and Chlorine of the rules.

The **metal** always stands **on the left**. Lithium has 1 **electron to give away**. That makes Lithium a metal, and he stands on the left.

The **non-metal** always stands **on the right**. Chlorine needs to **take in an electron** to have a complete Outside Energy Level. That makes him a non-metal, and Chlorine stands on the right.

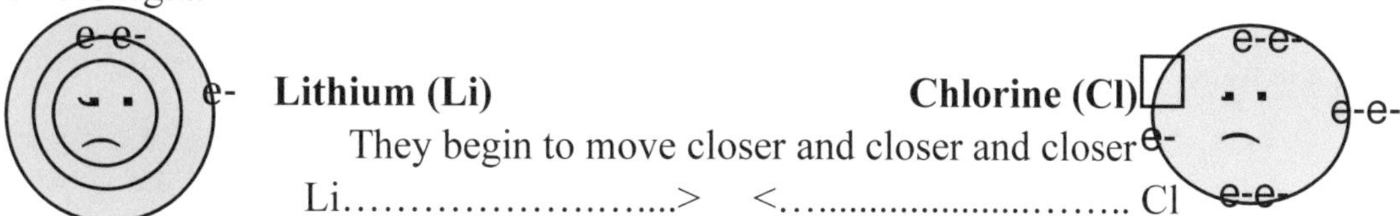

Lithium (Li) **Chlorine (Cl)**

They begin to move closer and closer and closer

Li……………………...> <…......................…….. Cl

Suddenly, when they get really close, Lithium's electron jumps into the empty space in Chlorine's Outside Energy Level. The moment this happens, there is a breathtaking sound. It's magical! As Lithium's electron moves into Chlorine's Outside Energy Level, Lithium's hidden energy level that is complete, pops up. Both atoms become COMPLETE. Look at their beautiful smiling faces! Notice Lithium became a happy atom with only 2 electrons. That's because he had only 1 energy level next to the nucleus, and it was complete with only 2 electrons.

Professor Terry said,"Looks like you've become a Happy Atom, Lithium!"

They were more than just happy. Now together the atoms were a new creation. They were no longer just Lithium and Chlorine. They were now a COMPOUND with a new name and formula.

Their new name is..............................Lithium Chlor**ide.**

Their formula is.......................................LiCl.

The non-metal element was the one to change the ending of his name,

*ine changed to ide, Chlor**<u>ine</u>** became Chlor**<u>ide</u>***

Here's how we write what happened in words and diagrams.

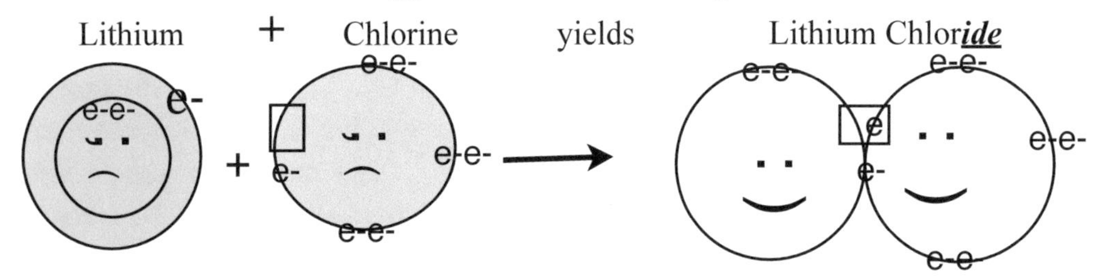

"Just like all the other compounds formed by these two families, it took only one atom of Lithium to make Chlorine's Outside Energy Level complete. To create the formula, we write their symbols together remembering that the metal goes first. The formula is **Li**Cl."

"Well Guy, what do you think?"

Guy said, "I'm glad you saved Lithium for last or I might not have noticed how Lithium was different. Thank you for teaching me all about Lithium."

"Glad you learned how Lithium was different. Yet he formed compounds using the same process." Professor Terry continued, "You have learned the secret of how the elements in Periodic Table Land become Happy Atoms. Watching the Alkali Metals form compounds with the Halogens is only the beginning of the Happy Atom Story. Elements are creative and they will find many ways to become Happy Atoms. You will get a chance to see how they do it.

This was the end of a very happy day. Guy and Professor Terry left Periodic Table Land happy. Professor Terry went to her lab. Guy went to his family's cabin on the mountain. Before the Alkali Metals left they said to Chlorine, "Let's duplicate these compounds. That way each family can have all of the Happy Atoms we made"

Chlorine said, "I'm so glad you thought of doing that." Chlorine took his magic wand wand and waved it over all the compounds they had made giving the Halogens duplicates of all the compounds. This made all the elements on 7th Street happier than they had ever been before.

Then the Alkali Metals went back to 1st Street with all the compounds they had made. Everyone of them were smiling the biggest smiles you have ever seen.

e

Sodium, delighted that the Alkali Metals had learned to be Happy Atoms, went on to visit the Alkaline Earth Metal family to share the good news.

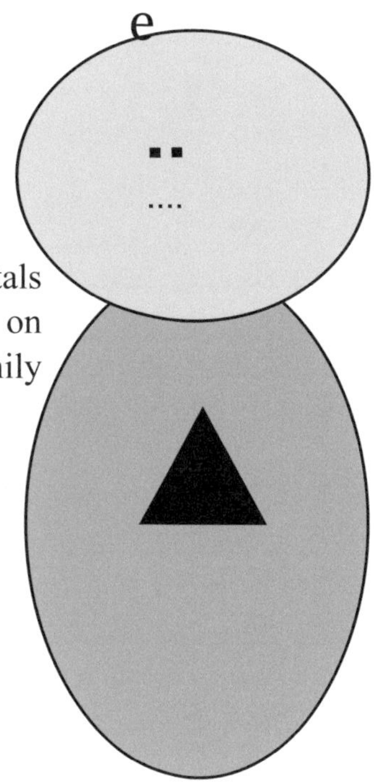

The Alkaline Earth Metals and The Oxygens Become Happy Atoms

Chapter 3

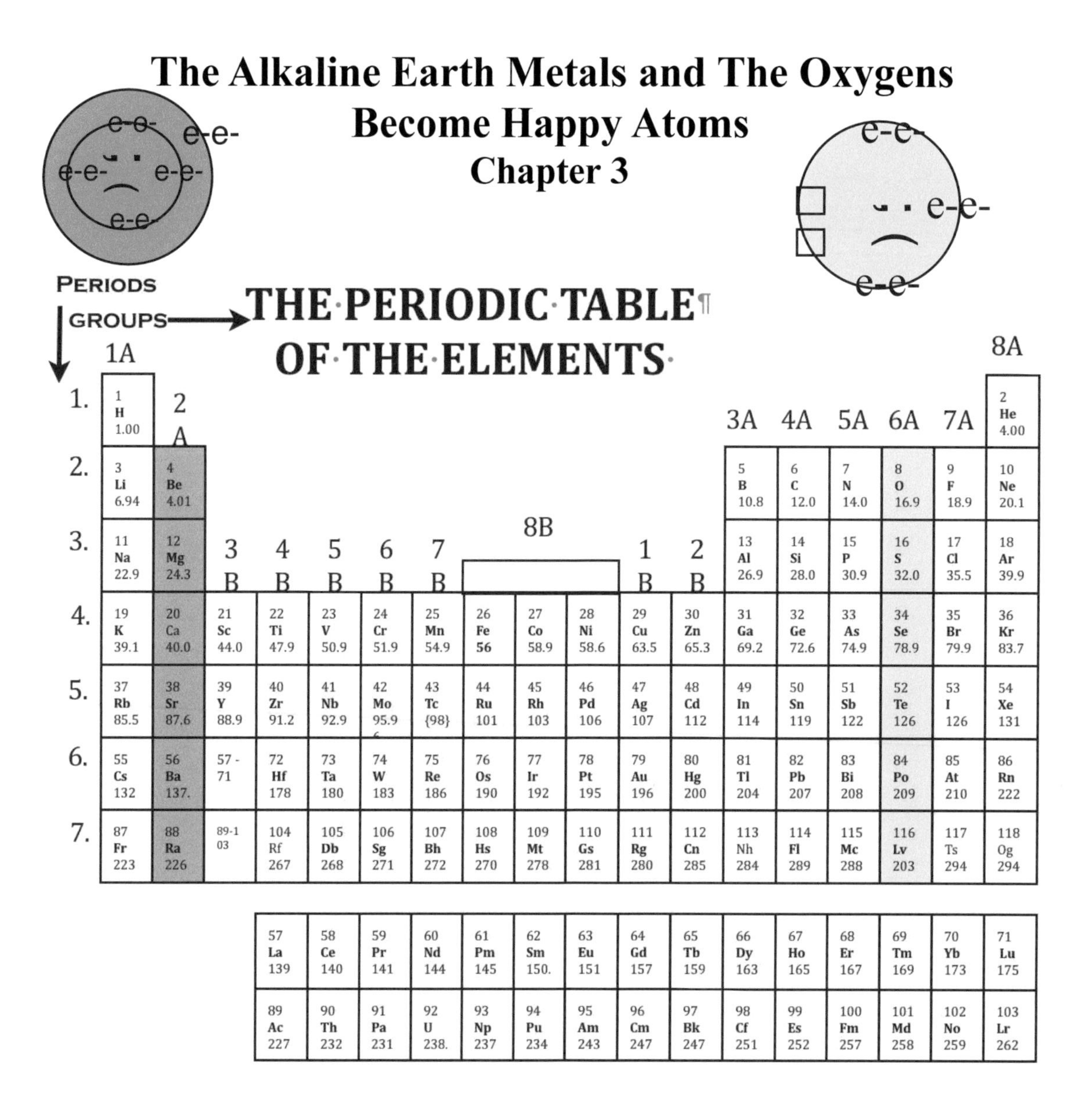

Periods / Groups	1A	2A	3B	4B	5B	6B	7B	8B	8B	8B	1B	2B	3A	4A	5A	6A	7A	8A
1.	1 H 1.00																	2 He 4.00
2.	3 Li 6.94	4 Be 4.01											5 B 10.8	6 C 12.0	7 N 14.0	8 O 16.9	9 F 18.9	10 Ne 20.1
3.	11 Na 22.9	12 Mg 24.3											13 Al 26.9	14 Si 28.0	15 P 30.9	16 S 32.0	17 Cl 35.5	18 Ar 39.9
4.	19 K 39.1	20 Ca 40.0	21 Sc 44.0	22 Ti 47.9	23 V 50.9	24 Cr 51.9	25 Mn 54.9	26 Fe 56	27 Co 58.9	28 Ni 58.6	29 Cu 63.5	30 Zn 65.3	31 Ga 69.2	32 Ge 72.6	33 As 74.9	34 Se 78.9	35 Br 79.9	36 Kr 83.7
5.	37 Rb 85.5	38 Sr 87.6	39 Y 88.9	40 Zr 91.2	41 Nb 92.9	42 Mo 95.9	43 Tc {98}	44 Ru 101	45 Rh 103	46 Pd 106	47 Ag 107	48 Cd 112	49 In 114	50 Sn 119	51 Sb 122	52 Te 126	53 I 126	54 Xe 131
6.	55 Cs 132	56 Ba 137.	57 - 71	72 Hf 178	73 Ta 180	74 W 183	75 Re 186	76 Os 190	77 Ir 192	78 Pt 195	79 Au 196	80 Hg 200	81 Tl 204	82 Pb 207	83 Bi 208	84 Po 209	85 At 210	86 Rn 222
7.	87 Fr 223	88 Ra 226	89-103	104 Rf 267	105 Db 268	106 Sg 271	107 Bh 272	108 Hs 270	109 Mt 278	110 Gs 281	111 Rg 280	112 Cn 285	113 Nh 284	114 Fl 289	115 Mc 288	116 Lv 203	117 Ts 294	118 Og 294

57 La 139	58 Ce 140	59 Pr 141	60 Nd 144	61 Pm 145	62 Sm 150.	63 Eu 151	64 Gd 157	65 Tb 159	66 Dy 163	67 Ho 165	68 Er 167	69 Tm 169	70 Yb 173	71 Lu 175
89 Ac 227	90 Th 232	91 Pa 231	92 U 238.	93 Np 237	94 Pu 234	95 Am 243	96 Cm 247	97 Bk 247	98 Cf 251	99 Es 252	100 Fm 257	101 Md 258	102 No 259	103 Lr 262

Magnesium, one of the Alkaline Earth Metals on 2nd Street, decided to visit Calcium another member of his family. He had an interesting observation. "I've been hearing laughter all morning coming from 1st Street. I bet they have learned how to be happy." Calcium and Magnesium were hoping that Sodium had found out how to be happy because it meant so much to everyone. When Magnesium left, Calcium sat dreaming of all the possible ways life could be better if they could only become happy.

Suddenly he was jarred out of his dream by a knock at the door. It was Sodium confirming what Magnesium had suspected. "I've come to tell you that my whole family has become happy. They are all now Happy Atoms. Do you remember when I set out to find a happy atom, I promised you I'd come back and tell you about my findings? Well, I'm here with the good news you've been waiting for."

Calcium never did beat around the bush. He came right out and said, "How did your family get to be happy?"

Sodium said, "I'd rather tell how *you* can become happy. It was when I finally got to 8th Street that I found atoms that were happy. Before that, there were no happy atoms anywhere. After I hung around 8th Street for a while, I learned that they were happy because all of them had complete Outside Energy Levels. They were the only atoms in Periodic Table Land that were happy, and the only ones that had complete Outside Energy Levels. None of the other families in all of Periodic Table Land had complete Outside Energy Levels, and all of them were just as sad as your family and mine."

"So, believing a complete Outside Energy Level was the reason an atom is happy made sense. The question was how to get that complete energy level and become happy. My family became happy by getting rid of the one electron that was in our Outside Energy Level. When we got rid of that electron, our complete energy level that was hidden popped up and became our complete Outside Energy Level. What you need to remember is this, the secret of becoming a Happy Atom is to achieve a complete Outside Energy Level."

"I have an idea for your family that I'm reasonably sure will work because it worked for my whole family," continued Sodium. "Let's look at your atom. You are in Group 2A, so all the elements in your family have 2 electrons in their Outside Energy Level. It takes 8 electrons to be complete. Let's look at your options. You could take in 6 more electrons which would give you the 8 electrons needed to become complete. However, the easier way is to give away those 2 electrons**.** The energy level below will then become your Outside Energy Level. Take a peek and see that hidden energy level is already a complete energy level."

Complete Energy Level

Calcium looked deep into his atom and saw that the hidden energy level had the 8 electrons in it that was required to be complete. "That sounds like the best idea," said Calcium. Calcium began to be excited about the possibility of becoming a Happy Atom. "Tell me how I can get rid of these two electrons."

"Listen carefully," said Sodium. "The Oxygen family elements on 6th Street have 6 electrons in their Outside Energy Level. Take a look at this picture of Oxygen's Outside Energy Level. He needs 2 more electrons to have a complete Outside Energy Level."

Calcium saw the 2 empty spaces where his 2 electrons could go.

"Oxygen will be happy to get your 2 electrons," said Sodium. "When he gets those electrons he will have the 8 electrons he needs to have a complete Outside Energy Level and become a Happy Atom. You can bring your whole family to 6th Street, and you can all become Happy Atoms the way we did on 7th Street with the Halogens. I'm sure you will be welcomed by the Oxygen Family."

Sodium volunteered to go along with the Alkaline Earth Metals and be the director since he knew the rules that needed to be followed. This was comforting to Calcium. Its's always good to have an experienced person with you when you try something for the first time.

Calcium gathered together the rest of the Alkaline Earth Metals: Magnesium, Beryllium, Strontium, Barium and Radium. Then, Sodium explained what would happen. When they understood, they got packed for the trip.

Sodium went to see Professor Terry and Guy and explained his plan for having the Alkaline Earth Metals become Happy Atoms. Sodium invited them to watch.

Professor Terry and Guy flew to 6th Street in their comfortable magic bubble. They set up an observation post where they assumed the action would take place.

In the meantime Sodium joined Calcium and the rest of the Alkaline Earth Metals: Magnesium, Beryllium, Strontium, Barium and Radium. They all slid into the bubble that Calcium's magic wand had enlarged to be bus size. Sodium admired the earthy color of the Alkaline Earth Metal balloons, so different from his family's light green balloons. He waved his magic wand and pointed it in the direction of 6th Street. Soon they arrived at their destination and walked over to Oxygen's lovely home. The fresh air made them breathe much better.

When they got to Oxygen's home, Sodium explained everything to Oxygen. Sodium told him, "You are in Group 6A. So you have 6 electrons in your Outside Energy Level. You need 8 to become a Happy Atom. So you only need to get 2 more electrons to be complete. Each Alkaline Earth Metal has the 2 electrons you need."

Oxygen said he understood. Then he shared some important facts about his family. "Sulfur is a non-metal like me and a close friend. The other members of my family are not friendly. I know Sulfur will be as interested as I am in becoming a Happy Atom, but we'll forget about the rest of the family for now. Maybe someday they will be interested." So, Sodium, Calcium and Oxygen went to see Sulfur.

Because of Sodium's excellent teaching, Oxygen had a good understanding of what needed to take place to become a Happy Atom and was able to explain it to Sulfur. Oxygen and Sulfur were ready to form compounds and become happy.

Sodium, the director, got everyone organized. Calcium turned to Oxygen and Sulfur and said, "I have 5 more Alkaline Earth Metal elements with me who want to become Happy Atoms. I'd like to introduce them to you both. So Oxygen and Sulfur meet my family: Magnesium, Beryllium, Strontium, Barium and Radium."

Sodium selected Calcium and Oxygen to go first. Sulfur and the rest of the Alkaline Earth Metals moved over to the side of the grassy area to watch how the process worked. Their job was to figure out how this was done. Then they would know what to do when it was their turn.

Professor Terry and Guy moved their magic bubble where they could observe the Alkaline Earth Metals making compounds and becoming Happy Atoms. They heard

Sulfur sigh, “I’m glad I’m not going first.” They were soon in position ready to watch the action.

Calcium and Oxygen Become Happy Atoms

Sodium directing the process told Calcium and Oxygen to get ready to begin. “**Calcium**, you have electrons to give away. You’re a **metal**. You stand here on the **left**. **Oxygen,** you need to take in Calcium’s electrons. That makes you a **non-metal**. You stand over there on the **right**.”

“It’s simple. On 3 start moving toward each other. When you are close enough Calcium’s electrons will jump into your Outside Energy Level, Oxygen. On three begin 1, 2, 3…”

Ca ....>   <..O

So, Calcium and Oxygen moved toward each other. When they reached that certain closeness, there was that magical sound that Sodium remembered so well. He heard it at the exact moment Calcium’s 2 electrons moved into Oxygen’s Outside Energy Level. Look what happened! Calcium’s 2 electrons have filled in the empty spaces in Oxygen’s atom, and Calcium’s hidden complete energy popped up to make Calcium’s atom complete, too. Look how happy Calcium and Oxygen are now.

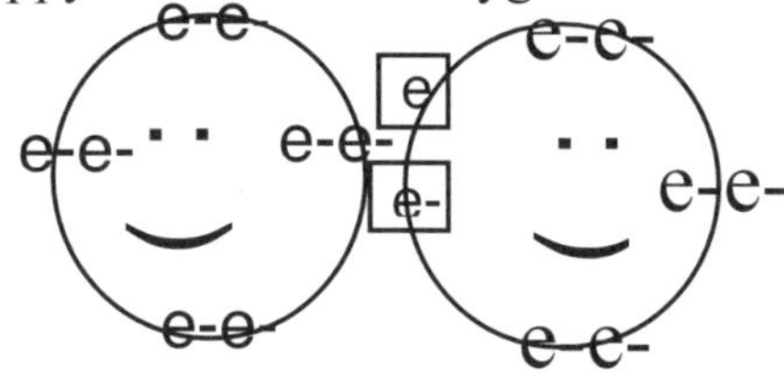

Calcium and Oxygen became a new compound, and they were happy. They have a new name, and their symbols joined together to create a formula.

This is their new name ....Calcium Ox**ide**

This is the new formula ………..CaO

*Notice Oxygen dropped **ygen** and changed it to **ide.***

Ox**ygen** changed his name to Ox**ide**

This change of name always happens when Oxygen forms a compound with one other element. This is what happened in words and a diagram.

Calcium   plus   Ox***<u>ygen</u>***   yields   Calcium Ox***<u>ide</u>***

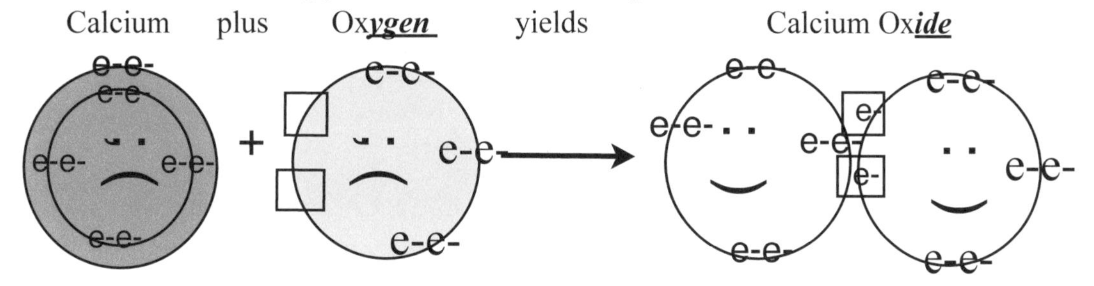

It took just **one Calcium atom** to provide **one Oxygen atom** the 2 electrons it needed to have a complete Outside Energy Level and become a compound. So we just wrote the two symbols together as the formula: **CaO.**

A Formula represents the elements that have joined together, using the symbols for each element, and it tells how many atoms of each element joined. Each symbol represents 1 atom of that element. **CaO** says it only took one atom of Calcium and one atom of Oxygen to form this compound, **Calcium Oxide.**

All the rest of the elements in the Alkaline Earth metal family have two electrons to give away. Oxygen needs exactly 2 electrons to make his Outside Energy Level complete. So they all form compounds in exactly the same way that Calcium did. Each of the elements took their turn, making compounds with Oxygen one after the other. All the compounds were formed with one atom of Oxygen and one atom of the Alkaline Earth metals. Here's how we write it in words and in diagrams.

The Alkaline Earth Metals combine with Oxygen to Form Compounds

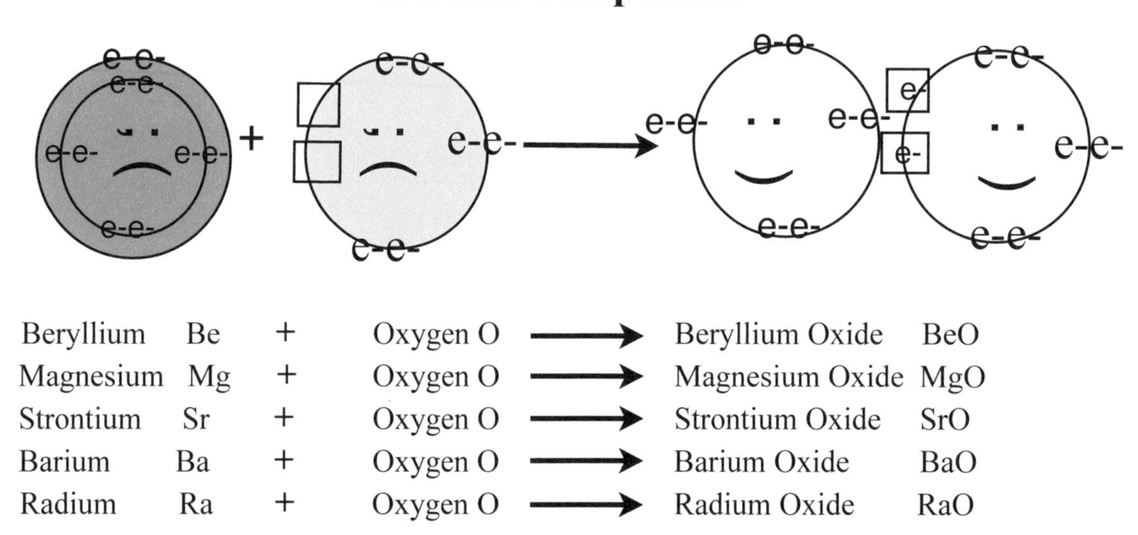

Beryllium	Be	+	Oxygen O	⟶	Beryllium Oxide	BeO
Magnesium	Mg	+	Oxygen O	⟶	Magnesium Oxide	MgO
Strontium	Sr	+	Oxygen O	⟶	Strontium Oxide	SrO
Barium	Ba	+	Oxygen O	⟶	Barium Oxide	BaO
Radium	Ra	+	Oxygen O	⟶	Radium Oxide	RaO

And they all became Happy Atoms!

Professor Terry turned to Guy and said, "Wasn't it wonderful to see how the Alkaline Earth Metals became Happy Atoms with Oxygen? Now we are going to see how they become Happy Atoms with Sulfur, another element in the Oxygen Family."

Magnesium and Sulfur Form A Compound

Sodium still the director, turned his attention to Sulfur. "I guess you know the routine now. I noticed you were watching carefully as Oxygen formed compounds with all the Alkaline Earth Metals. Do you feel ready Sulfur? The time to line up is now."

Magnesium, the **Metal** element with the electrons to give away stood on the **left**.
Sulfur the **non-metal** element that needed to get the electrons stood on the **right**

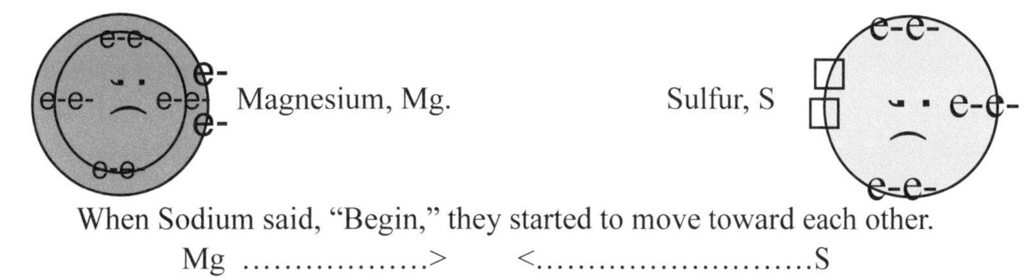

When Sodium said, "Begin," they started to move toward each other.
Mg ………………> <………………………S

When they reached that special distance apart, Magnesium's two electrons jumped into the empty spaces in Sulfur's Outside Energy Level, and that famous sound happened! Here's what they looked like. There's Sulfur with Magnesium's 2 electrons filling his 2 empty spaces. There's Magnesium with his hidden energy level popped up. His Outside Energy Level is now complete. They both are Happy Atoms. See them smiling."

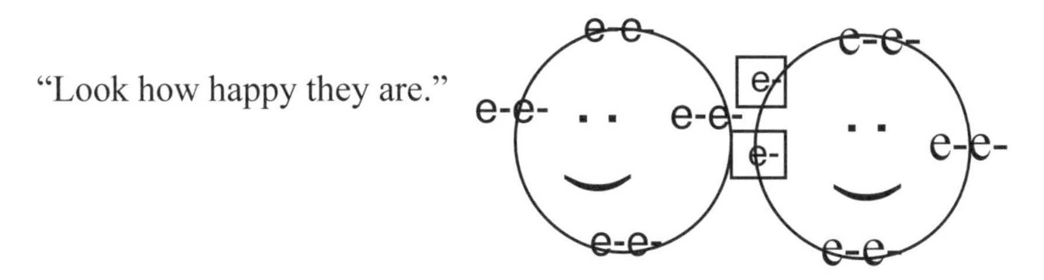

They have become a compound with a new name and formula.
They have a **new name** ....Magnesium Sulf***ide*.**
Their symbols changed into a **formula** ....MgS
Notice Sulf**ur** dropped the ending **ur** and added **ide.**
*So Sulf**ur** became Sulf**ide**.*
Remember, it's the non-metal element that needs to change his name's ending

This is how we write what happened in words and diagrams.

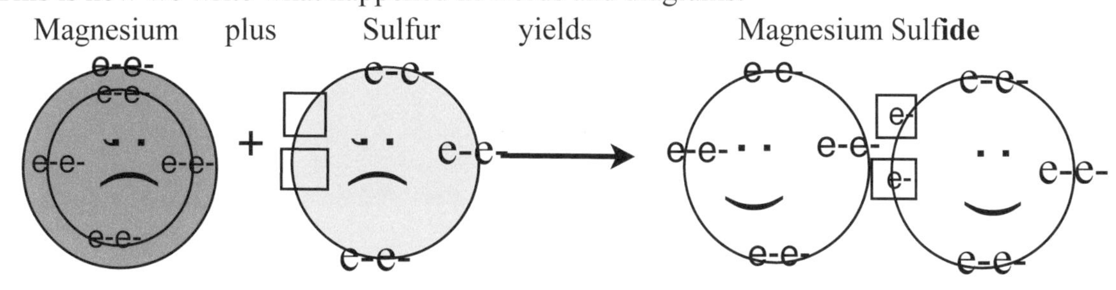

It only took one atom of Magnesium to provide Sulfur with the 2 electrons he needed to have a complete Outside Energy Level. Only one atom of each element joined to form this compound. Because of this, we just write the two symbols together to make the formula, Magnesium Sulfide, MgS. This formula says it took 1 atom of Magnesium and 1 atom of Sulfur to form the compound, Magnesium Sulfide, MgS.

Guy Shows He Understands Compound Formation

Guy whispered to Professor Terry, "I noticed again the element on the right changed the ending his name to ***ide.*** This time Sulfur dropped the ***ur*** before adding the ending *ide* and became Sulf**ide**. I remember Oxygen dropped "***ygen***" before adding ***ide*** and became Ox**ide.**"

She responded, "You are going to have to remember this Guy. When only two elements join together to form a compound, the non-metal changes the ending of his name to ***ide.*** Oxygen compounds are Ox**ides**. Sulfur compounds are Sulf**ides.**

"OK Guy, I want you to understand this. All the Alkaline Earth Metal elements form compounds and become a Happy Atom with Sulfur in the exact same way that Magnesium formed a compound with Sulfur. For each one you can look at the steps Magnesium went through to form a compound and all do it the same way."

Guy said, "I get it. Let me guess what will happen when Barium tries to become Happy Atoms with Sulfur."

Professor Terry said, "If you can show me how Barium makes a compound with Sulfur, I'll be sure you understand the process."

With confidence Guy began. "Here's Barium combining with Sulfur. It's easy."
Barium stands on the left because he is the **metal.** He has electrons to give away.
Sulfur, the **non metal** stands on the right because he needs to get electrons.

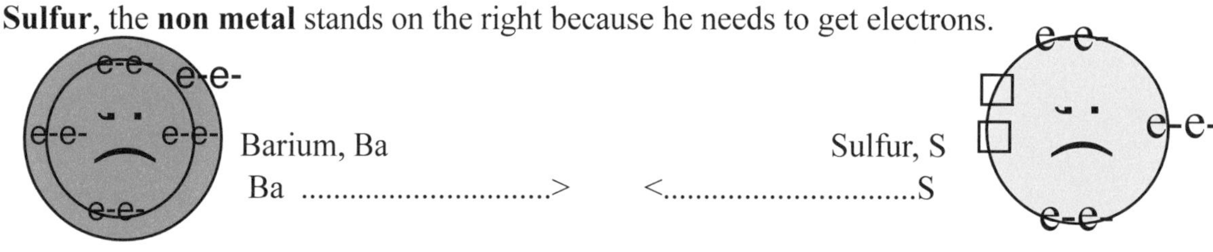

They move closer and closer. Then Barium's 2 electrons jump into Sulfur's Outside Energy Level, and they get to be a new compound with smiling happy faces.

They are Happy Atoms and become a compound.
They have a new name.........,,,Barium Sulf***ide.***
Their symbols changed to a formula...BaS
Guy *took off Sulfur's ending, **ur** and changed it to **ide***
*Sulf**ur** became Sulf**ide.***

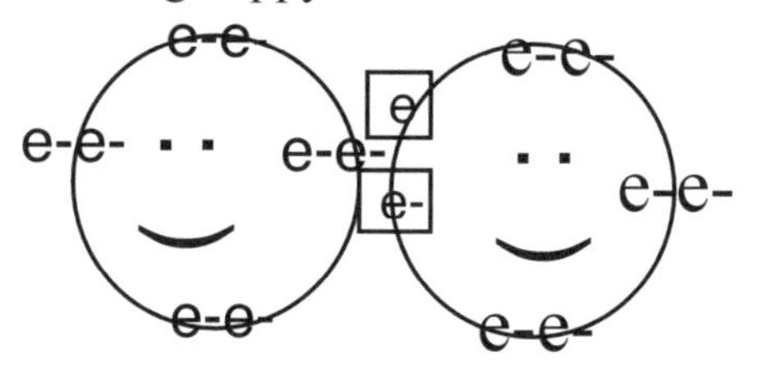

Now I'll write what happened in words and a diagram.

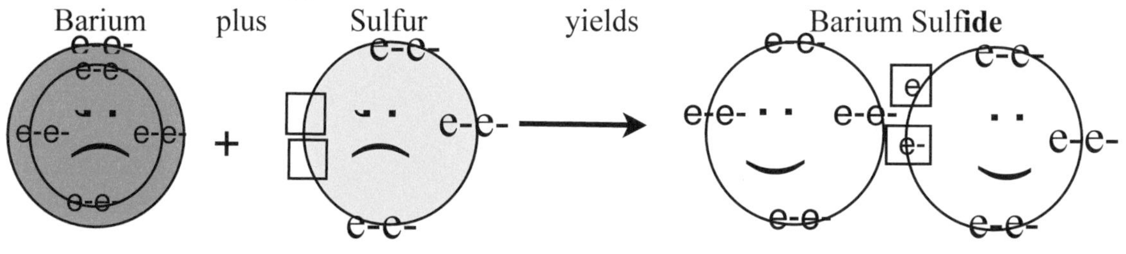

Professor Terry congratulated Guy for knowing how Barium became a Happy Atom and how well he wrote their new name and formula. All the rest of the Alkaline Earth Metals form compounds in the same way. Here's the list."

The Alkaline Earth Metals Combine With Sulfur to Form Compounds

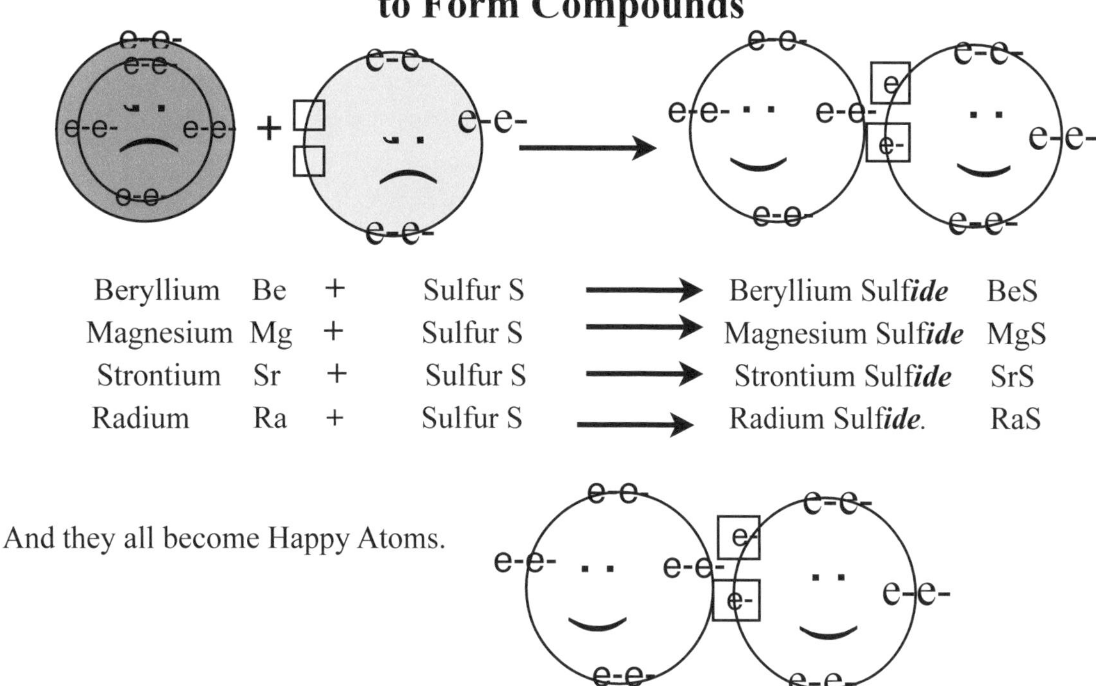

Beryllium	Be	+	Sulfur S	→	Beryllium Sulf***ide***	BeS
Magnesium	Mg	+	Sulfur S	→	Magnesium Sulf***ide***	MgS
Strontium	Sr	+	Sulfur S	→	Strontium Sulf***ide***	SrS
Radium	Ra	+	Sulfur S	→	Radium Sulf***ide***.	RaS

And they all become Happy Atoms.

Beryllium Becomes A Happy Atom with Sulfur

Professor Terry began telling Guy about a special element in the Alkaline Earth Metal family. "Beryllium is like Lithium in the Alkali Metal Family. They both are left with only one energy level after they give away the electrons in their Outside Energy Level. I think you remember from drawing Bohr models that the first energy level next to the nucleus is complete with only 2 electrons. So, Beryllium like Lithium is left with a complete energy level that has only 2 electrons in it. That means Beryllium's compound is going to look different from all the other elements in the Alkaline Earth Metal family."

"Here's what Beryllium's energy levels look like."

"There will be only one more metal in Periodic Table Land like this. That element is Boron in Group 3A. Lithium, Beryllium and Boron can become Happy Atoms with only 2 electrons in their Outside Energy Level because two electrons make the energy level next to the nucleus complete. I would like you to see how this happens when Beryllium forms a compound with Oxygen."

"Here's how Beryllium and Oxygen become Happy Atoms." Sodium reminded Beryllium and Oxygen of the rules."

"The **metal** always stands **on the left**. Beryllium has 2 **electrons to give away**. That makes him a metal, and he stands on the left."
The **non-metal** always stands **on the right**. Oxygen needs to **take in 2 electron** to have a complete Outside Energy Level. So Oxygen is a non-metal, and he stands on the right."

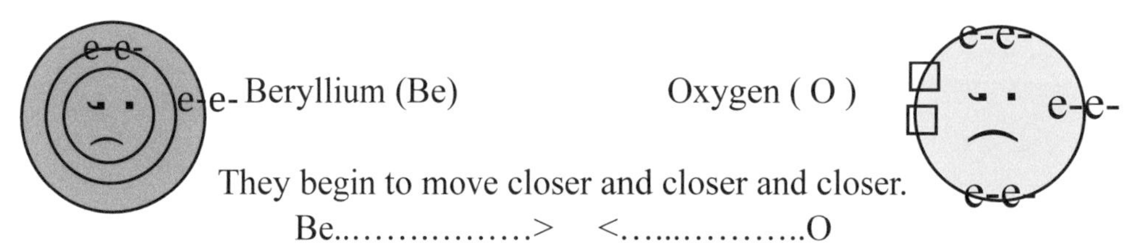

They begin to move closer and closer and closer.

Be.................> <.................O

Suddenly, when they got really close, Beryllium's 2 electrons jumped into the two empty spaces in Oxygen's Outside Energy Level. The moment this happened, there was a breathtaking sound. It was magical! As Beryllium's 2 electrons moved into Oxygen's Outside Energy Level, Beryllium's hidden energy level, that was complete, popped up. Both atoms became COMPLETE. Look at their beautiful smiling faces! Notice that Beryllium's Happy Atom has only 2 electrons not 8 like the rest of the Alkaline Earth Metals when they formed a compound. Remember that's because Beryllium's energy level was the one next to the nucleus which is complete with only 2 electrons.

They are now Happy Atoms.

They were more than just happy. Now together the atoms were a new creation. They were no longer just Beryllium and Oxygen. They were now a COMPOUND with a new name and formula.

Their new name is..................................Beryllium Ox***ide.***

Their formula is...BeO.

The non-metal element was the one to change the ending of his name.

Oxygen** dropped ygen and changed to ide, **Oxygen** became Ox**__ide__

Here's how we write what happened in words and diagrams.

Beryllium + Oxygen yields Beryllium Ox***__ide__***

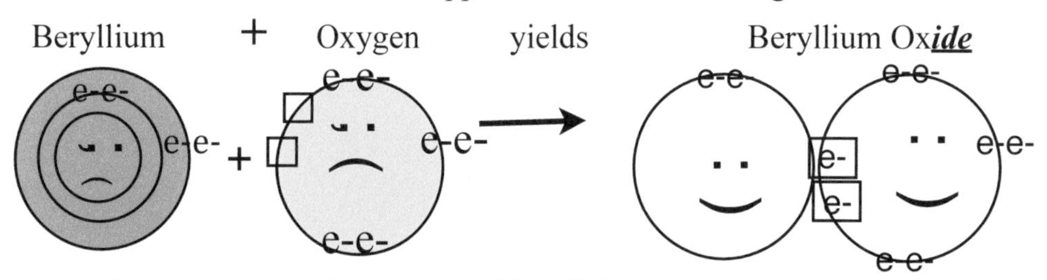

Professor Terry continued, "Just like all the other compounds formed by these two families, it took only one atom of Beryllium to make 1 atom of Oxygen's Outside Energy Level complete. To create the formula, we just write their symbols together remembering that the metal goes first. The formula is **BeO**."

The Alkaline Earth Metal and Oxygen Families Celebrate

Sodium gathered Oxygen, Sulfur and all the Alkaline Earth Metals together. Sodium said, "Now it's time to celebrate." They all went down to the meadow. with Oxygen and Sulfur leading the way. Strontium one of the Alkaline Earth Metals lit off the fireworks. After that, they built a bonfire and roasted marshmallows. Then they sang and danced around the fire to celebrate having learned how to form compounds—how to become Happy Atoms. They certainly were happy, Happy Atoms.

After the excitement of the celebration quieted down, Professor Terry and Guy began to talk. Guy said, "Watching the elements make Happy Atoms was very interesting. Group 1A became happy when they found a family that needed one electron. Group 2A elements became happy when they found a family that needed 2 electrons. What's next?"

Just then Guy saw Aluminum running back toward 3rd Street. Guy asked, "Professor Terry did you notice Aluminum hanging around while the Alkaline Earth Metals were forming compounds?"

Professor Terry responded, "I think he was there to witness all the Alkaline Earth Metals becoming Happy Atoms. He's probably thinking of forming Happy Atoms too. We'll find out what's going on soon enough."

"You asked what was next. Going to 3rd Street was in my plan even before I saw Aluminum checking things out. In fact Sodium has already gone to 5th Street trying to prepare the way for the 3rd Street Boron family to make compounds with the Nitrogen family. He's presently there explaining how they can become Happy Atoms with the elements on 3rd Street. Group 5A elements are missing 3 electrons and the elements on 3rd Street have the 3 electrons they need to give away in order to get a complete Outside Energy Level. So 5th Street is where Aluminum should go to become a Happy Atom."

The celebration ended, and Professor Terry and Guy set out for 3rd Street. Professor Terry waved her magic wand and their bubble rose straight up and over the trees. They were on their way to see what was going on over at 3rd Street. While they were high above the trees, Guy asked Professor Terry another question that bothered him. "Why did Aluminum come to observe Group 2A atoms become happy rather than Boron, the leader of 3rd Street?"

Professor Terry explained, "3rd Street and 4th Street are special neighborhoods. The elements in these families have many differences. Some act as metals, some are metalloids. I think you remember that metalloids sometimes act as metals and other times as non-metals. The elements in Group 2A are all the same. They are all metals. Boron, the leader of 3rd Street, is a metalloid and in addition he doesn't really like giving away his electrons the way metals do. So to answer your question, Guy, Boron knew that Aluminum, was the best one for the job. Aluminum was a metal and most like the Alkaline Earth Metals in Group 2A. So who would better understand what was going on?

Guy answered, "Aluminum, a metal, would probably understand better what the metals were doing to form compounds."

Professor Terry concluded, "Boron, like all good leaders, just sent the best person for the job."

The Boron and Nitrogen Families Become Happy Atoms

Chapter 4

THE PERIODIC TABLE OF THE ELEMENTS

PERIODS ↓ GROUPS →

Period	1A	2A	3B	4B	5B	6B	7B	8B			1B	2B	3A	4A	5A	6A	7A	8A
1.	1 H 1.00																	2 He 4.00
2.	3 Li 6.94	4 Be 4.01											5 B 10.8	6 C 12.0	7 N 14.0	8 O 16.9	9 F 18.9	10 Ne 20.1
3.	11 Na 22.9	12 Mg 24.3											13 Al 26.9	14 Si 28.0	15 P 30.9	16 S 32.0	17 Cl 35.5	18 Ar 39.9
4.	19 K 39.1	20 Ca 40.0	21 Sc 44.0	22 Ti 47.9	23 V 50.9	24 Cr 51.9	25 Mn 54.9	26 Fe 56	27 Co 58.9	28 Ni 58.6	29 Cu 63.5	30 Zn 65.3	31 Ga 69.2	32 Ge 72.6	33 As 74.9	34 Se 78.9	35 Br 79.9	36 Kr 83.7
5.	37 Rb 85.5	38 Sr 87.6	39 Y 88.9	40 Zr 91.2	41 Nb 92.9	42 Mo 95.9	43 Tc {98}	44 Ru 101	45 Rh 103	46 Pd 106	47 Ag 107	48 Cd 112	49 In 114	50 Sn 119	51 Sb 122	52 Te 126	53 I 126	54 Xe 131
6.	55 Cs 132	56 Ba 137.	57 - 71	72 Hf 178	73 Ta 180	74 W 183	75 Re 186	76 Os 190	77 Ir 192	78 Pt 195	79 Au 196	80 Hg 200	81 Tl 204	82 Pb 207	83 Bi 208	84 Po 209	85 At 210	86 Rn 222
7.	87 Fr 223	88 Ra 226	89-103	104 Rf 267	105 Db 268	106 Sg 271	107 Bh 272	108 Hs 270	109 Mt 278	110 Gs 281	111 Rg 280	112 Cn 285	113 Nh 284	114 Fl 289	115 Mc 288	116 Lv 203	117 Ts 294	118 Og 294

57 La 139	58 Ce 140	59 Pr 141	60 Nd 144	61 Pm 145	62 Sm 150.	63 Eu 151	64 Gd 157	65 Tb 159	66 Dy 163	67 Ho 165	68 Er 167	69 Tm 169	70 Yb 173	71 Lu 175
89 Ac 227	90 Th 232	91 Pa 231	92 U 238.	93 Np 237	94 Pu 234	95 Am 243	96 Cm 247	97 Bk 247	98 Cf 251	99 Es 252	100 Fm 257	101 Md 258	102 No 259	103 Lr 262

While Professor Terry and Guy were on their way to 3rd Street, Aluminum had already arrived back there out of breath. He sat on the bench in front of his unique house, made of twisted, flat, and crumpled aluminum, all configured in a tasteful way. The brass plate above the door with the number #13 engraved on it announced that this was the home of Aluminum, Atomic Number 13, his home.

He had been on 6th Street and watched the Alkaline Earth Metals become Happy Atoms with the Oxygen family. He thought about the events he had witnessed. Watching the Alkaline Earth Metals become Happy Atoms with elements in the Oxygen Family was an awesome experience. When Aluminum figured out what he would report to Boron, he walked slowly to house #5, 3rd Street to see Boron, the head of his family. As he walked there he was thinking about how much nicer it would be on 3rd Street if they could become Happy Atoms. Aluminum found Boron napping in his hammock under the sweet smelling magnolia tree beside his home. After poking Boron awake, he shared the fantastic events he observed on 6th Street. He explained how all the Alkaline Earth Metals formed compounds with the Oxygen Family, and all of them became Happy Atoms. “It’s so exciting because it gives us hope that we can be happy too.”

Aluminum had hardly finished telling Boron what he had observed, when Professor Terry and Guy brought their bubble down behind 3rd Street brushing the leaves on the upper branches as they landed. They could hear Aluminum and Boron talking. Professor Terry and Guy followed the voices and found Boron and Aluminum out in the garden. The fragrance of the Magnolia blossoms filled the air.

Professor Terry said, hardly pausing to breathe, “Hi, Boron and Aluminum. Exciting things have been happening around Periodic Table Land. Everyone is learning how to become Happy Atoms. It’s all about your electrons. To be specific, it’s about the electrons in your Outside Energy Level. That energy level needs to be complete for an element to be happy. How many electrons do you have in your Outside Energy Level?”

Boron said, “Everyone in my family has 3 electrons there. Three is the number of electrons in our Outside Energy Level, because we live on 3rd Street, that’s Group 3A on the Periodic Table.”

Professor Terry continued her lesson saying,“Take a look at Aluminum’s energy levels. Guy, you remember the short way to display energy levels without drawing a Bohr model, don’t you?”

Guy said, “I do. Aluminum’s atomic number is 13. His energy levels are:

2) 8) **3**).”

Boron said, “My Atomic Number is 5. Here’s what my energy levels look like:

2) **3**).”

“All of us on 3rd Street, as I said before, have 3 electrons in our Outside Energy Level. That’s how we are alike. But look, how really different we are. Notice all the electrons Aluminum has. He has a whole energy level of electrons more than I have. That’s 8 electrons more than I have. Maybe that’s why I don’t feel like giving away my electrons?”

Professor Terry smiled. “I think you’re on to something but it is more complicated than just not feeling like giving away electrons. Your 3 electrons want to stay because the protons in your nucleus do not want them to go. Remember protons are positively charged and are attracted to the electrons with negative charges. This attractive force keeps the electrons close to the nucleus. It is easier for Aluminum to give away his 3 electrons because the protons still have 10 electrons left to keep his 13 protons happy. Aluminum’s protons will not miss the 3 electrons that leave. When you, Boron, give away your 3 electrons, your protons are left with only 2 electrons. You are probably right to believe that having only 2 electrons is not enough to keep your protons happy. They will miss the 3 that leave. That’s a hypothesis—a reasonable guess as to the cause of a situation. Boron, I promise you, we’ll find a way that you can use your electrons without making your protons unhappy, but first, I’d like to work on Aluminum becoming a compound.

Aluminum, Al

Here’s a picture of Aluminum’s last 2 energy levels, showing the 3 electrons he has to give away.”

Nitrogen, N

“Look! Do you see those 3 empty spaces in Nitrogen’s Outside EnergyLevel he needs to fill.?”

Boron said, "I see that Aluminum has 8 electrons hidden under his energy level with 3 electrons in it. The 8 electrons make the hidden level complete."

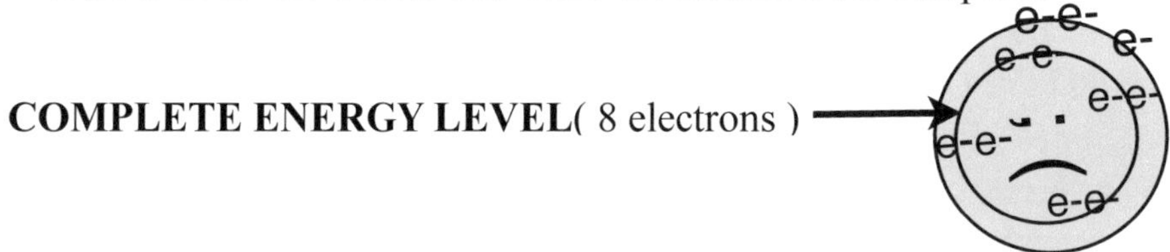

Now Professor Terry said. "You are on the right track! If Aluminum didn't have those 3 electrons in the outside level, his complete energy level would be his Outside Energy Level. Then he would become a Happy Atom because his Outside Energy Level would be complete."

Boron said, "It looks like I have another problem to think about. My hidden energy level only has two electrons in it. My energy levels look like this 2) 3).

When I get rid of my 3 outside electrons, I'll only have 2 electrons left, not 8 like Aluminum. I'm worried. It looks like I won't become a Happy Atom after all."

Guy knew what to tell Boron. "I'm happy to tell you, Boron, that the energy level next to the nucleus is complete with only 2 electrons. If you get rid of your 3 electrons, you will have only one energy level left. When there is only one energy level next to the nucleus, it is complete with 2 electrons. You will be happy because you will have a complete Outside Energy Level."

Professor Terry went on to assure Boron, "Don't worry Boron. Guy's right, when you have only one energy level, 2 electrons makes it complete. The only thing that matters is that the Outside Energy Level must be complete. Boron, I saw Lithium and Beryllium become Happy Atoms and their energy levels were just like yours. After they gave away their outside electrons, they were left with one energy level. **It was complete with 2 electrons**, and they became Happy Atoms. They formed a compound and you will too."

Boron was relieved to hear this good news. He would have a chance to become a Happy Atom, after all. "I must find a way to make my protons happy."

Aluminum, hardly able to contain himself, exclaimed, "I was on 6th Street when Group 2A and 6A elements formed compounds and became Happy Atoms. I'm pretty sure I know that we can become Happy Atoms too. On the way back from watching the Oxygen family become happy, I noticed that the elements on 5th Street have 5 electrons in their Outside Energy Level. That means that they need 3 more electrons to have the 8 electrons needed to be complete. Our 3 electrons will give them a complete Outside Energy Level. I took a picture, so I could show you

"If we go to 5th Street and explain how to become a Happy Atom, I bet the atoms in the Nitrogen Family will take our 3 outside electrons right away to fill those 3 empty spaces and become complete. I know they are tired of being so sad. I say we go to 5th Street as fast as possible, and try to get Nitrogen to make Happy Atoms with us."

As Professor Terry was leaving, she said to Aluminum, "It sounds like you know how to become a Happy Atom. I'm glad, but if there is any little detail you are not sure of, Guy and I will be on 5th Street to help you. See you there."

Aluminum called together all the members of the Boron Family and told them the story of how to become happy. Gallium, Indium and Thallium were not that interested. They were Post Transition Metals and had different ideas. "I guess they don't mind being sad, but we want to be happy. It will be just the two of us going off on this adventure," said Aluminum. "These three elements are missing an opportunity to be happy. I guess they will just wait until they're ready. Let's get over to 5th Street."

Their family bubble lifted by those lovely silver balloons landed them behind the forest at the head of 5th Street. Silver was Boron's family color, and Boron was proud of it. Boron hung back at the edge of the forest, while Aluminum approached Nitrogen with his plan. Phosphorus happened to be visiting Nitrogen at that time. So, Aluminum explained the whole deal to both of them at the same time. They quickly understood the concept of acquiring a complete Outside Energy Level.

Nitrogen checking to see if he understood correctly said, "Let me get this straight. If I take your three electrons, both you and I will become happy. Am I right?"

"That's right," said Aluminum. "If I give you my 3 electrons, you will have the 8 electrons you need to have a complete Outside Energy Level. When I get rid of those 3 electrons, my hidden complete energy level will become my Outside Energy Level. We will both become Happy Atoms and form a new compound."

Professor Terry and Guy were there but decided to keep out of the way and just observe. They moved their bubble to the edge of the tree line with a good view of the clearing on 5th Street. Here's what they saw happening.

Aluminum and Nitrogen Become Happy Atoms

Aluminum, the **metal** with 3 electrons to give away, stood on the left.
Nitrogen, the **non-metal** element that needed to get the 3 electrons, stood on the right.

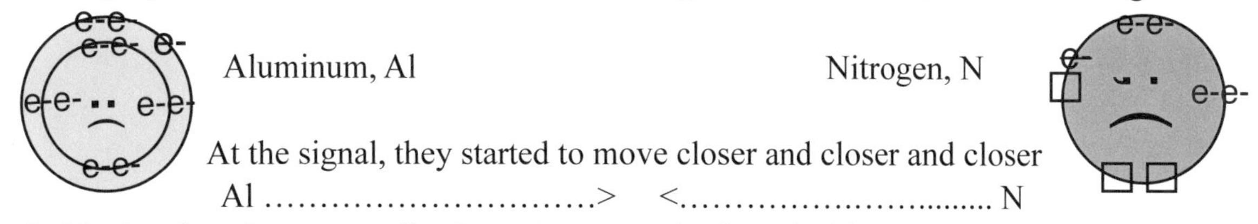

Suddenly when they got really close, there was that breathtaking sound which marked the exact moment when Aluminum and Nitrogen joined together and became Happy Atoms. It was magical! Aluminum's hidden energy level that was complete, popped up making Aluminum happy. Nitrogen got Aluminum's 3 electrons and became complete. Both atoms being *complete* formed a compound.

Look at their smiling happy faces!
They were truly Happy Atoms.

They were more than just happy. They were a beautiful new creation, a compound.

They have a **new name**. ....Aluminum Nitr**ide**

Their symbols changed into a formula Al N

Notice the non-metal was the one to change his name.

*Nitrogen dropped his ending **ogen** and added **ide***

*Nitr**ogen** became Nitr**ide**.*

Since it took only one atom of Aluminum to give one atom of Nitrogen the 3 electrons he needed, you just write the two symbols together as the formula, Al N

This is how you write what happened in words and diagrams.

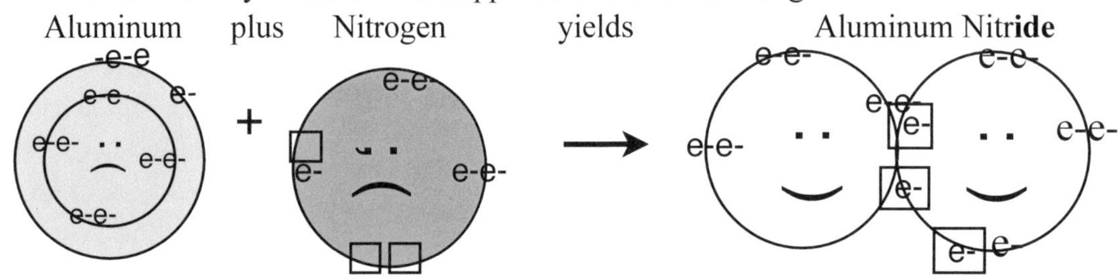

Professor Terry assured Guy that he would learn how to write Chemical Equations like a scientist, with symbols and formulas, when the time was right.

Aluminum and Phosphorus Become HappyAtoms

Well, that was a breakthrough. Aluminum had become a Happy Atom with Nitrogen. Boron and Phosphorus were observers. They were encouraged that it looked possible for them to become happy too. However, they had their concerns. Aluminum said, "I'll show you that I can become a Happy Atom with Phosphorus too. That way, Boron, you'll be sure this will work the same way with Phosphorus just as well as with Nitrogen." So Aluminum and Phosphorus lined up in the proper way.

Aluminum, the element with 3 electrons to give away, **the metal** stood on the left.
Phosphorus, the element that was missing 3 electrons, **the non-metal** who needed to get the electrons stood on the right.

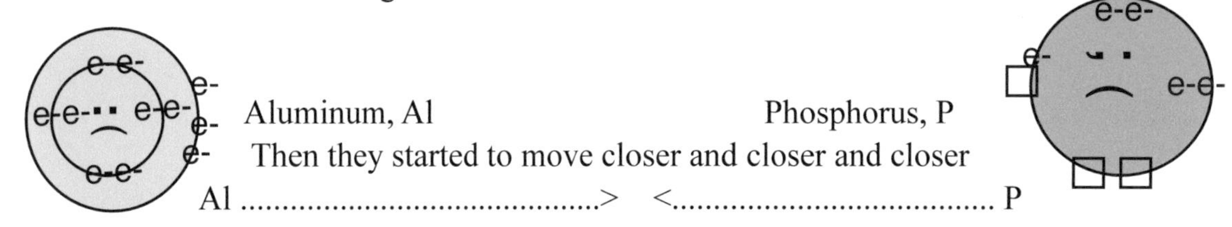

Suddenly, when they got really close, there was that breathtaking sound proclaiming the momentous event when Aluminum's 3 electrons moved into the 3 empty spaces in Phosphorus' Outside Energy Level giving Phosphorus the 8 electrons he needed to be complete and happy. Watch what happened! Aluminum's hidden energy level that was complete, popped up making Aluminum happy. Both atoms became *complete,* formed a compound and they became Happy Atoms.

Look at their beautiful smiling faces.

They were more than just happy.
They were something new— a compound.
They were a beautiful new creation and very Happy Atoms.

They have a **new name** ....Aluminum Phosph**ide**

Their symbols changed to a **formula..................**AlP

Notice the element Phosphorus that took in the extra electrons, the non-metal
is written on the right in the compound's name and formula
He changed the ending of his name from ***orus*** *to* ***ide***
*Phosph****orus*** *became Phosph****ide****.*

*S*ince it took only one atom of Aluminum to give Phosphorus the 3 electrons he needed you just write the two symbols together as the formula, AlP.

This is how you write what happened in words and diagrams.

Aluminum plus Phosphorus yields Aluminum Phosphide

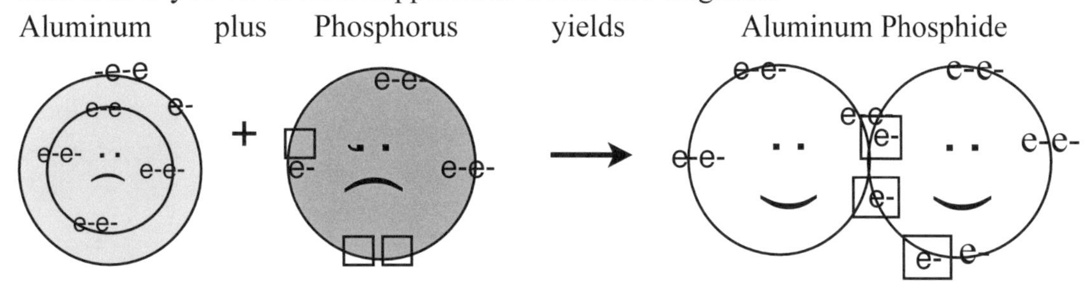

After watching Aluminum make Happy Atoms with Nitrogen and Phosphorus, Boron knew that the process of forming compounds and becoming a Happy Atom was simple enough; but something was bothering Boron. He turned to Aluminum for support and said, "I want to become a Happy Atom but I don't want to make my protons unhappy. To become a Happy Atom, I have to give away my 3 outside electrons but my protons don't want me to give them away. The protons have grown accustomed to having these electrons close. The positive protons have a force that attracts the negative electrons and holds them close to the nucleus. This force is like the force of a magnet keeping the electrons near. This is their job and they will be very sad if I do give their electrons away."

Aluminum said, "My protons weren't sad when I gave away my 3 electrons."

"Well Aluminum, when you gave away your 3 electrons your protons still had 10 electrons to pull on to keep near the nucleus. When I give away my 3 electrons my protons will be left with only 2 electrons, that's less than half of the electrons they are used to having close to them and that's not enough. They will be so sad."

"Now you know my problem. How can I become a Happy Atom and still keep my protons happy?"

Aluminum wanted to find a way to make Boron happy. After all he was the leader of 3rd Street. Aluminum thought of the many ways elements tried to be happy before they learned to become Happy Atoms. Aluminum said, "I was relatively happy when I found myself in a sheet of aluminum foil or in an aluminum pan." He tried to think of something that could make Boron happy but couldn't come up with even one good idea.

Professor Terry from the vantage point of her bubble saw that Aluminum and Boron were in need of help. So she and Guy climbed out of the bubble and crossed the lawn to see how to help. A while ago, Boron had shared with Professor Terry his concern about making his protons sad if he chose to be a HappyAtom. So Professor Terry had time to think of a solution for Boron.

Boron said,"Oh I'm so glad you are here. You promised to find a special way for me to become a Happy Atom that will not make my protons sad. "

Professor Terry looked confident saying, "When I was thinking of a solution for you, I remembered how Oxygen became happy. It gave me an idea. In the air, Oxygen managed to be happy when 2 Oxygen atoms learned to share electrons. Maybe you can share your electrons with other atoms instead of giving electrons away and still become happy. When you share electrons instead of giving them away, it's called covalent bonding. Think of covalent as cooperating. Sharing electrons is like cooperating. If you are willing, let's try sharing electrons with Nitrogen. It might be a way for you to become a Happy Atom and keep your protons happy too. Here's the part of my idea that you will appreciate. When you share your electrons with Nitrogen, your protons will now have Nitrogen as a next door neighbor. Your protons will be sharing Nitrogen's electrons the way neighbors share lawns in the common area between houses. So your protons will have even more electron friends than they had before."

Boron wanting so much to become a Happy Atom agreed to try. "It sounds like a great idea, and I'm willing to do it. However, I must say that I need to be brave because I am just trusting that you are right about everything. It's scary but worth trying."

Aluminum, Guy and Professor Terry in one happy chorus said, "You'll do fine Boron. Stop worrying. You are very brave, and we know you'll be happy that you tried."

"OK," said Boron, "I'm ready to begin now. We'll soon find out."

Boron and Nitrogen Become Happy Atoms

Boron, a metalloid, acting as a metal, with electrons to get rid of, stood on the left.
Nitrogen, the **non-metal,** the element that needed to get the electrons stood on the right.

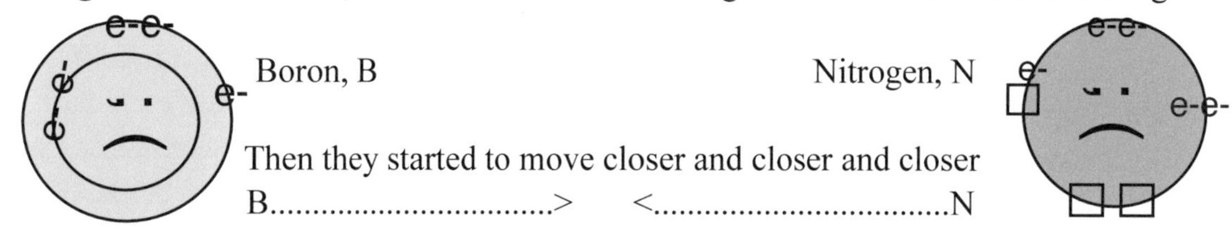

Suddenly, when they got really close, Guy heard that unique sound again proclaiming that Boron's 3 electrons had indeed been pulled into the 3 empty spaces in Nitrogen's Outside Energy Level. This gave Nitrogen the 8 electrons he needed to be complete and happy. In that same instant Boron's hidden energy level was freed up becoming Boron's complete Outside Energy Level. Boron was happy. Boron and Nitrogen became a new compound. They were Happy Atoms at last. Boron and Nitrogen looked at each other and said, "Did you notice we are sharing my 3 electrons? I guess being covalent worked. The main thing is we are Happy Atoms, and we made a new compound."

Boron said, "Also, my dear little protons did not have to lose their buddies, the electrons.

Look at our smiling happy faces!

Boron said, "I'm amazed. I'm different in so many ways, I'm a metalloid. My Outside Energy Level has only 2 electrons in it. I'm sharing electrons to make my protons happy. How wonderful! I created a compound and became a Happy Atom!"

Nitrogen and I have a **new name** **................**Boron Nitr**ide**
Our symbols changed into a **formula................ BN**

Notice *the non-metal that took in the extra electrons is written on the right in the compound and the formula. It is the non-metal that changed the ending of his name.*

*Nitr**ogen** dropped **ogen,** and added **ide .***
*Nitr**ogen** became Nitr**ide.***

Since it took only one atom of Boron to give Nitrogen the 3 electrons he needed, you just write the two symbols together as the formula——**BN**. The formula BN says that 1 atom of Boron combined with 1 atom of Nitrogen to form the compound Boron Nitride.

This is how you write what happened in words and diagrams.

Boron Plus Nitrogen yields Boron Nitride

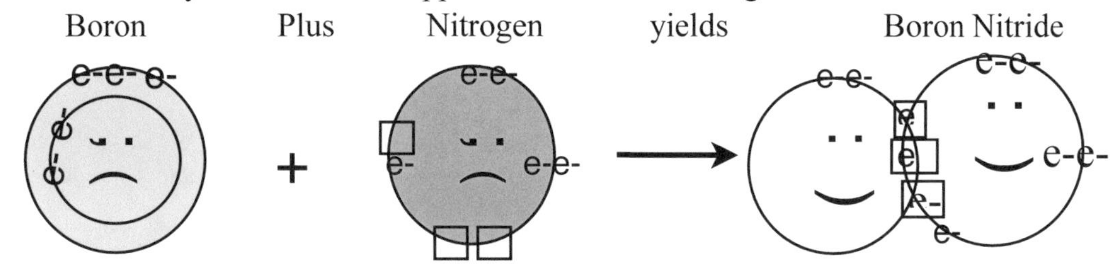

Boron and Phosphorus Become Happy Atoms

Boron, not worried any more said, "Let's not waste any time I'm ready to see if it's possible to create a new compound with Phosphorus. So they lined up and said, "Let the action begin."

Boron, a metalloid, acting as a **metal** with electrons to give away stood on the left.
Phosphorus, a non-metal, the element that needed to get electrons, stood on the right.

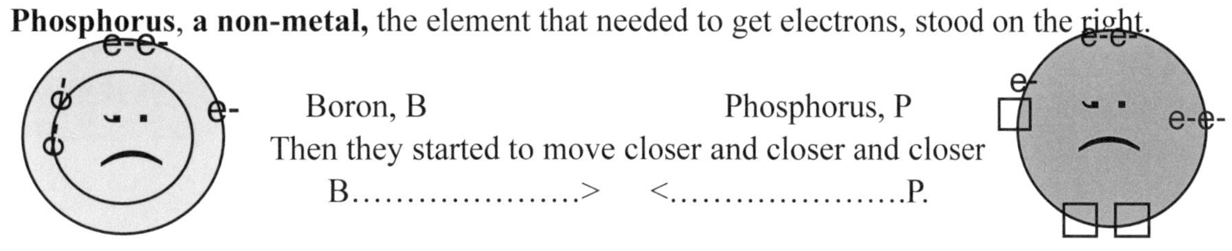

Boron, B Phosphorus, P
Then they started to move closer and closer and closer
B....................> <.....................P.

Suddenly, when they got really close, there was that famous sound again that happened precisely as Boron's 3 electrons were pulled into the 3 empty spaces in Phosphorus' Outside Energy Level giving Phosphorus the 8 electrons he needed to be complete. Boron's hidden energy level, that was complete with only 2 electrons, popped up, and they both became Happy Atoms. Nitrogen and Phosphorus said at the same time, "Did you notice we are sharing electrons? We are covalent and I'm complete with only 2 electrons in my Outside Energy Level. Notice too my protons are happy that I didn't give away their electron buddies. They even get to share your electrons as neighbors. This is so amazing!"

Look at our big smiles!

We became something new and wonderful, a compound.

We have a **new name**................................Boron Phosph**ide**

Our symbols changed into a **formula** ....BP

*Notice, the element that took in the extra electrons, the non-metal, Phosphorus is written on the right in the compound and the ending of his name is changed. Phosph**orus** dropped the **orus**. And added **ide.** Phosph**orus** became Phosph**ide.***

Since it took only one atom of Boron to give Phosphorus the 3 electrons he needed, you just write the two symbols together as the formula for Boron Phosphide, **BP.**

This is how you write what happened in words and diagrams.

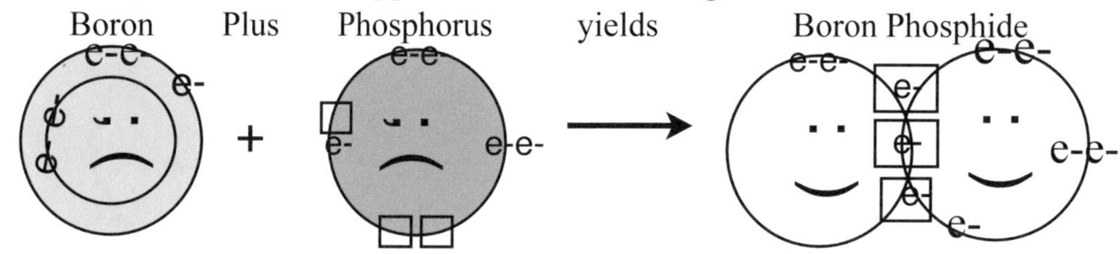

Professor Terry turned to Guy and said, "Aluminum, Boron, Nitrogen and Phosphorus all became Happy Atoms. We made Boron's protons happy and Boron was able to become a Happy Atom."

Guy ask, "Professor Terry, I know that Boron became a compound by sharing electrons. Could you tell me a little more about how this worked?"

Professor Terry explained, "It was simple. Boron, the metal, just stretched his Outside Energy Level over near the non-metal's Outside Energy Level. The non-metal's force pulled these electrons into Nitrogen's empty spaces while Boron's electrons still remained in Boron's energy level. This means the protons were able to remain close to their dear little electron friends. Boron was happy that he did not have to give his electrons away and make his protons sad. A compound was formed, the protons remained happy. Best of all Boron and Nitrogen became Happy Atoms."

The Alkali Metals and Oxygen Family
Become Happy Atoms
Chapter 5

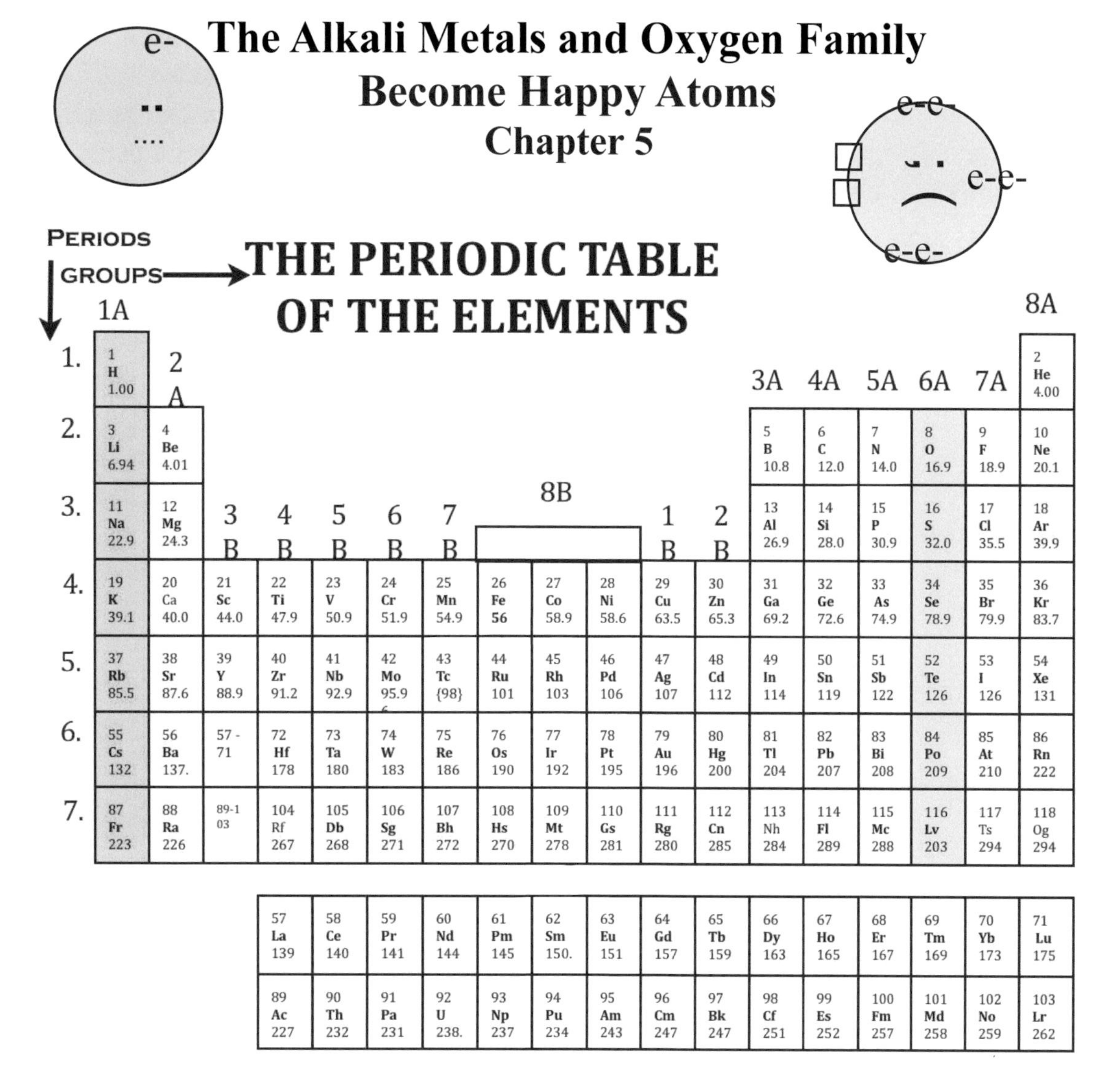

THE PERIODIC TABLE OF THE ELEMENTS

Period	1A	2A	3B	4B	5B	6B	7B	8B			1B	2B	3A	4A	5A	6A	7A	8A
1.	1 H 1.00																	2 He 4.00
2.	3 Li 6.94	4 Be 4.01											5 B 10.8	6 C 12.0	7 N 14.0	8 O 16.9	9 F 18.9	10 Ne 20.1
3.	11 Na 22.9	12 Mg 24.3											13 Al 26.9	14 Si 28.0	15 P 30.9	16 S 32.0	17 Cl 35.5	18 Ar 39.9
4.	19 K 39.1	20 Ca 40.0	21 Sc 44.0	22 Ti 47.9	23 V 50.9	24 Cr 51.9	25 Mn 54.9	26 Fe 56	27 Co 58.9	28 Ni 58.6	29 Cu 63.5	30 Zn 65.3	31 Ga 69.2	32 Ge 72.6	33 As 74.9	34 Se 78.9	35 Br 79.9	36 Kr 83.7
5.	37 Rb 85.5	38 Sr 87.6	39 Y 88.9	40 Zr 91.2	41 Nb 92.9	42 Mo 95.9	43 Tc {98}	44 Ru 101	45 Rh 103	46 Pd 106	47 Ag 107	48 Cd 112	49 In 114	50 Sn 119	51 Sb 122	52 Te 126	53 I 126	54 Xe 131
6.	55 Cs 132	56 Ba 137.	57 - 71	72 Hf 178	73 Ta 180	74 W 183	75 Re 186	76 Os 190	77 Ir 192	78 Pt 195	79 Au 196	80 Hg 200	81 Tl 204	82 Pb 207	83 Bi 208	84 Po 209	85 At 210	86 Rn 222
7.	87 Fr 223	88 Ra 226	89-1 03	104 Rf 267	105 Db 268	106 Sg 271	107 Bh 272	108 Hs 270	109 Mt 278	110 Gs 281	111 Rg 280	112 Cn 285	113 Nh 284	114 Fl 289	115 Mc 288	116 Lv 203	117 Ts 294	118 Og 294

57 La 139	58 Ce 140	59 Pr 141	60 Nd 144	61 Pm 145	62 Sm 150.	63 Eu 151	64 Gd 157	65 Tb 159	66 Dy 163	67 Ho 165	68 Er 167	69 Tm 169	70 Yb 173	71 Lu 175
89 Ac 227	90 Th 232	91 Pa 231	92 U 238.	93 Np 237	94 Pu 234	95 Am 243	96 Cm 247	97 Bk 247	98 Cf 251	99 Es 252	100 Fm 257	101 Md 258	102 No 259	103 Lr 262

Sodium was the hero of Periodic Table Land**.** By this time, most of the Chemical Families knew the way to become Happy Atoms because of Sodium. It was great to see so many Happy Atoms with big smiling faces walking around Periodic Table Land. Those who were not happy at least knew there was a way to become happy. Things were going along just fine.

Sodium was the one to apply the scientific method. Using investigation he was able to find a way to help all the Chemical Families become happy. In addition, he added to the scientific world by learning how to create new compounds.

Presently there was only one way for elements to create new compounds and become Happy Atoms. The only way was to find other elements that needed the exact same number of electrons as the elements had to give away. The elements on 1st Street, the Alkali Metals, with only one electron to give away found the Halogens on 7th Street who needed only one electron. The Alkaline Earth Metals on 2nd Street with 2 electrons to give away found the Oxygen Family on 6th Street who needed exactly 2 electrons. The Boron Family on 3rd Street with 3 electrons to give away found the Nitrogen Family on

5th Street who needed 3 electrons. The elements with electrons to give away had a hidden complete energy level which popped up after giving away those outside electrons. That's the way they achieved a complete Outside Energy Level. The elements that received those electrons took in the exact number of electrons they needed to have a complete Outside Energy Level. Then both atoms together became a new creation, a compound. They were Happy Atoms tucked away inside of these wonderful new compounds.

Stretched out on his comfortable lounge chair in his rose garden, Sodium contemplated the reason these elements became Happy Atoms. The peaceful sounds from a nearby stream and the sweet smell of the roses were conducive to creative thinking. There were no distractions in this peaceful garden. Thinking was Sodium's favorite pastime. So he thought and thought trying to find an idea that he liked. He said out loud, "I have to find a way for my family to become happy with elements that need more than one electron." Suddenly, an idea came to him. "I've got it! If it works, there will be many more Happy Atoms in Periodic Table Land." He was so excited!

Here's the idea that came to him. The Oxygen Family elements on 6th Street need two electrons to form a compound.

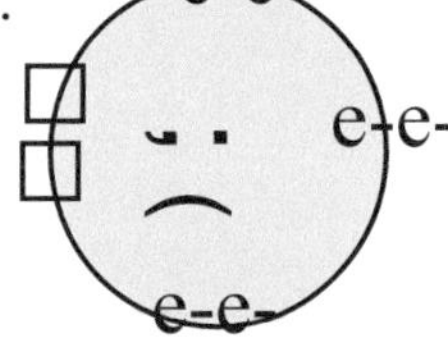

I wonder what would happen if two Sodium atoms

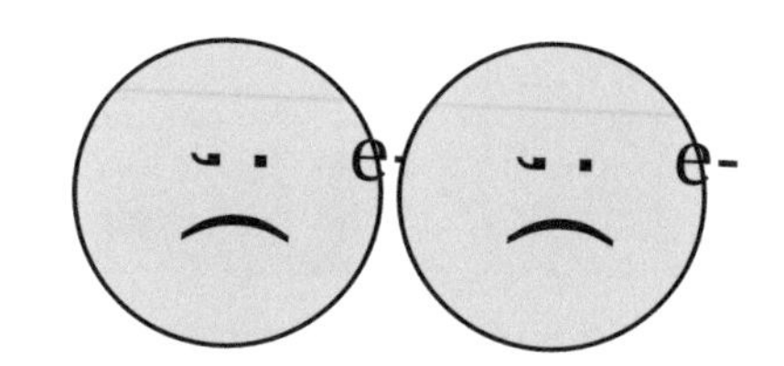

went to 6th Street, and each gives Oxygen one electron? Sodium continued to plan his strategy. He pondered out loud, "Oxygen would get the two electrons he needs to become a new compound and become a Happy Atom. Each Sodium atom would get rid of his one extra electron. Then both the Sodium atoms would become Happy Atoms too." It sounded reasonable to Sodium. He went over the scenario once more to make sure his thinking was logical. "That's right." he said out loud. "When each of the 2 Alkali Metals gives away his 1 electron, each of their hidden complete energy levels will pop up, and each atom will have a complete Outside Energy Level. Yes! This idea is good. Not just one, but two Sodium atoms will become Happy Atoms for every one Oxygen atom they join up with. This is very good."

"It's a great idea, and it sure sounds workable. But an idea is only a hypothesis," he reminded himself. "**A hypothesis is an idea about how something could reasonably happen**. Scientists need to experiment to prove that it can actually work." Sodium knew this. He said, "I guess I've got to start testing to see if my idea actually works."

Sodium made his way to the university lab to bounce his idea off Professor Terry, a scientist he respected. Guy was sitting on a tall stool by the lab table near Professor Terry when Sodium arrived. Guy was thrilled to listen as Sodium revealed his new plan.

Professor Terry liked the idea. She said, "I really believe it can work. Guy and I want to be present to witness this new experiment."

Sodium said, "If this experiment works on 6th Street, making compounds will probably work on on 5th Street as well. The elements there need 3 electrons. So three of my Alkali Metals could each give them an electron and three of us will become Happy Atoms. I think you'll enjoy being right up front watching it all happen. Guy will definitely learn first hand how compounds are formed."

Guy said, "We'll be the silent observers. Then I'll learn how compounds are made in each of these new ways.This will be the biggest adventure of my whole summer." Their flying bubble took them to the woods near 6th Street where they set up a tent, and stored their supplies.

Professor Terry said, "We'll be here a long time. Each day we will go where the action is. Today it's on 6th Street. Tomorrow 5th Street. Who knows where the action will be after that. Let's get over to the grassy area on 6th Street where the action will most likely take place any time now. We don't want to miss any of this."

Sodium went to visit Oxygen to get him on board with his new plan. Oxygen invited Sodium in to discuss how this new plan would work. They sat down, and Sodium launched into the explanation of his plan."Notice," said Sodium, "I have one electron to give away. You need two electrons to have a complete Outside Energy Level. My idea is to get another Sodium atom, and each of us will give you an electron. That will add up to the two electrons that you need. Together we will form a compound and all of us will become Happy Atoms."

Sodium and Oxygen Form A Compound and become Happy Atoms

Oxygen, a man of action, said, "Let's try it. I guess we line up as usual."
2 **Sodium** atoms, the **metals**, with electrons to give away. stand **on the left side.**
Oxygen, the **non-metal**, needing to get 2 electrons stands **on the right**.

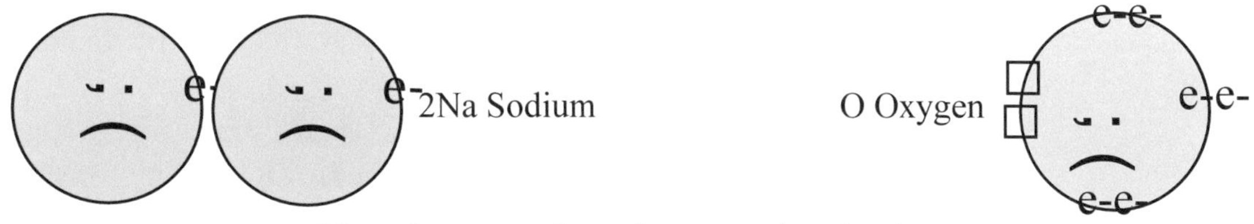

Then they started moving toward each other

2 Na ....> <...................................O

They finally reached that certain distance between them when the electrons jumped from each of Sodium's two atoms into the two empty spaces in Oxygen's Outside Energy Level. There was that magical sound that Sodium remembered so well, announcing the creation of a new compound. Each time it was a spectacular moment. This time it was two Sodium atoms and one Oxygen atom who became Happy Atoms.

They had hoped this would happen, but deep down they doubted it would. So this success was unexpectedly wonderful. Real possibilities for all the Alkali Metals to create new compounds were opened up. Oxygen knew Sulfur would also be delighted at the thought of making new compounds.

Let's see what actually happened. When each Sodium atom gave away his one electron, each got a complete Outside Energy Level. Each of their electrons moved into Oxygen's 2 empty spaces giving Oxygen a complete Outside Energy Level. All three atoms became Happy Atoms, and a new compound was formed. Look at them smile.

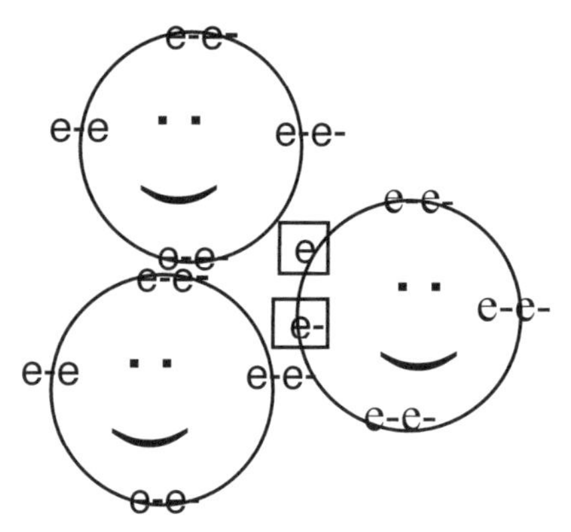

Not only are they Happy Atoms, but also another new compound was formed.

The **new name** is ....Sodium Oxide

The compound's **formula** is ....Na_2O

Naming the compound is done the same way as before.

The the non-metal is written on the right and changes its name.

***Oxygen** drops ygen, and adds ide.*

Ox***ygen*** becomes Ox***ide.***

Writing the formula for this new compound is different from the way we did it before because we now have more than one Sodium atom forming the compound. When there was only one atom of each element, we just wrote the symbols next to each other which meant it took one atom of each element to make the compound. When more than one atom of an element is involved, **the formula, uses subscripts to tell how many atoms of each element it took to make the compound.**

Professor Terry asked Guy, **"**How many Sodium atoms did it take to make Sodium Oxide? Count them."

He counted the Sodium atoms and said, "Two."

Moving on—Professor Terry asked, "How many Oxygen atoms did it take to make this compound?"

Guy responded, "One, and I remember when we write the formula we just write a symbol with no subscript. We never write 1 as a subscript because the symbol stands for 1 atom of an element. Writing the subscript one is not necessary."

Professor Terry complimented Guy for remembering that the symbol in a compound means there is one atom of that element in the compound. When there is more than one atom of an element in the compound, we write how many there are as a subscript. Therefore **the formula for Sodium Oxide is Na_2O.** We put a 2 under the Na because it took 2 atoms of Sodium to make this compound.

This is how we write what happened in words and diagrams.

2 atoms of Sodium plus 1 atom of Oxygen yields Sodium Oxide.

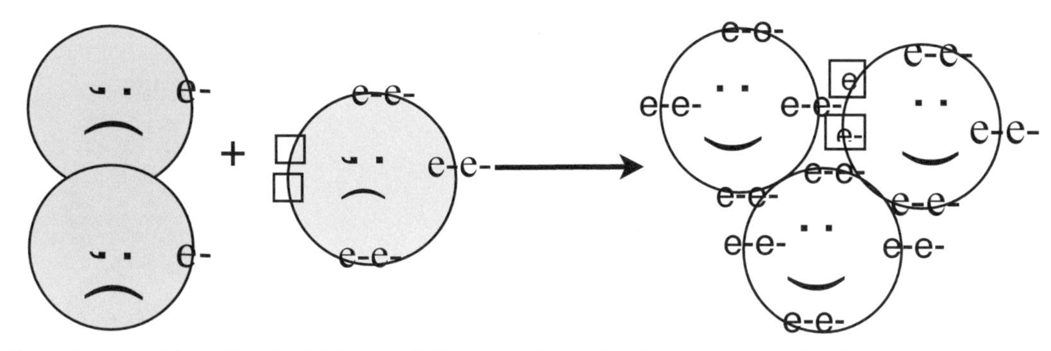

Sodium jumped into his bubble, and flew back to 1st Street, and told all the Alkali Metals the new way of becoming Happy Atoms on 6th Street. Potassium decided to come out with Sodium to see if he could make a new compound with a member of the Oxygen Family. He brought 2 atoms of Potassium with him. Oxygen told Sulfur, who needed 2 electrons, what to expect. So he was ready to begin. They immediately lined up to become Happy Atoms.

Potassium and Sulfur Form Compounds and Become Happy Atoms

2 atoms of **Potassium**, the metal, with electrons to give away, stood on the left. **Sulfur**, the non-metal, needing to get electrons, stood on the right.

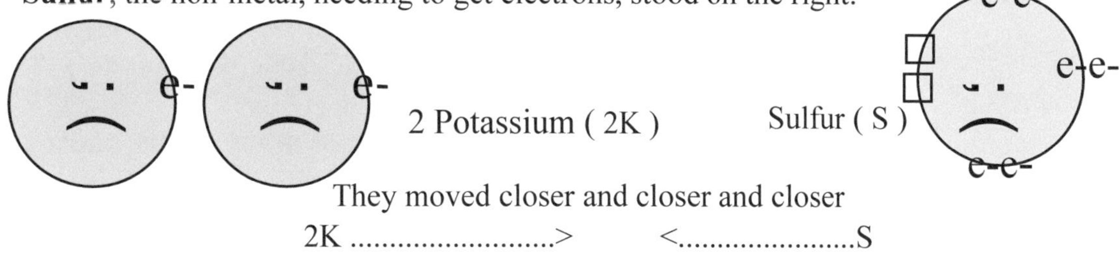

They moved closer and closer and closer

2K ....>   <......................S

When the atoms were finally close enough, the electrons from the two Potassium atoms jumped into the two empty spaces in Sulfur's Outside Energy Level. There was that magical sound that Potassium and Sulfur remembered so well announcing the creation of a new compound. Potassium and Sulfur were so very happy.

Look at their happy smiling faces.
They have a **new name**..............Potassium Sulf**ide.**
They now have a new **formula..**.........$K_2 S$

The subscript 2 means that it took 2 atoms of Potassium to combine with one atom of Sulfur to give them both complete Outside Energy Levels. Notice, the element that took in the extra electrons, the non-metal is written on the right in the compound and the ending of his name is changed.
*Sulf**ur** dropped the **ur**. And added **ide;** Sulf**ur** became Sulf**ide.***

From now on know that when Sulfur forms a compound with just one other element, the compound will always be called a Sulf**ide.**

This is how we write what happened in words and diagrams:

2 atoms of Potassium plus 1 atom of Sulfur yields Potassium Sulf**ide**

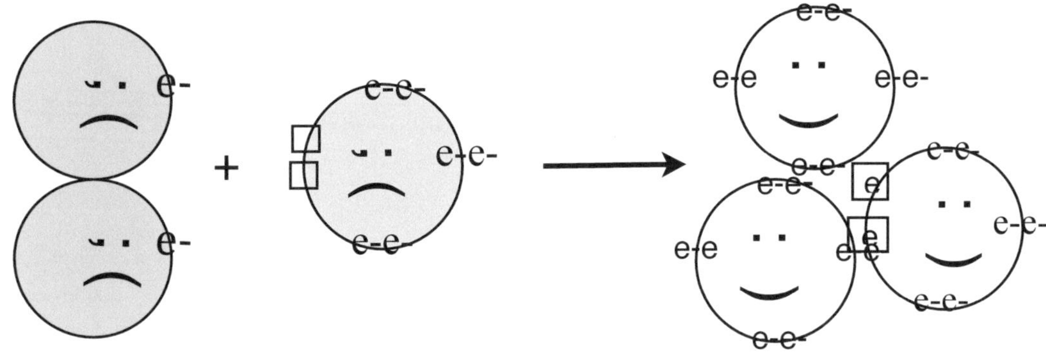

After that Sodium went back to 1st Street. He gathered his family around him and explained how he had found a new way to form compounds and become even happier Happy Atoms. Sodium said, “You have to go to 6th Street in pairs because the 6th Street elements need 2 electrons. So it takes 2 of you to come up with the two electrons that the elements in the Oxygen family need.”

They immediately understood and were up for the adventure. All the Alkali Metals piled into the flying family bus and flew out to 6th Street in no time. They were ready to create new compounds and become Happy Atoms. As they flew out there, they kept saying “The elements in the Oxygen Family need two electrons. We have only one electron to give away. So we have to remember to stay in pairs.” They were all ready to form compounds when they arrived.

Sodium, Potassium, Rubidium, Cesium, and Francium each took turns making compounds first with Oxygen and then with Sulfur.

“Here’s how they each formed a compound in words and diagrams.”

2 Alkali Metal elements plus 1 Oxygen Family element yields a Compound.

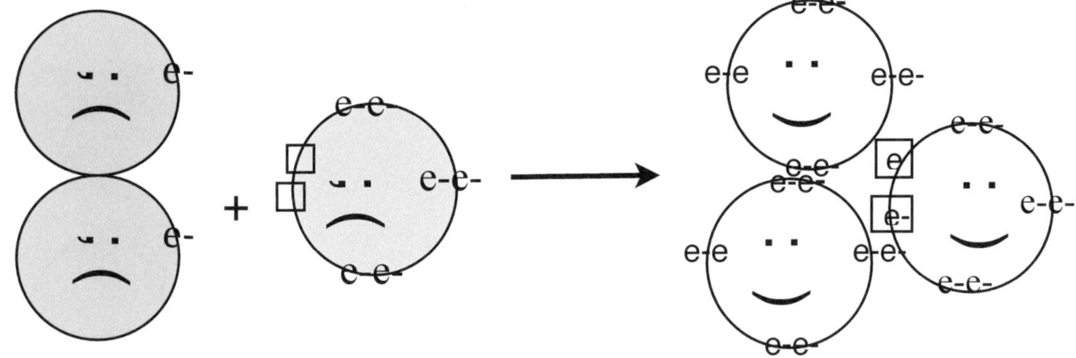

Just like this equation says, the elements went wild making Oxide and Sulfide compounds. Look at them all. All the compounds were made in the very same way. It took 2 Alkali Metals to provide the electrons needed by the elements in the Oxygen Family.

Look carefully at this diagram. This is how each of the Alkali Metals formed compounds with Oxygen or Sulfur to form Oxides or Sulfides. As they each became a Happy Atom a new compound was formed. The chart below shows each Alkali Metal combining first with Oxygen and then with Sulfur. The compound formed has a new name and a formula. Notice the subscript under each of the Alkali Metals in each compound is 2, and remember that it means it took 2 Alkali metals to combine with one Oxygen atom or one Sulfur atom to make the compound.

OXIDES		SULFIDES	
Sodium Oxide	Na_2O	Sodium Sulfide	Na_2S
Potassium Oxide	K_2O	Potassium Sulfide	K_2S
Rubidium Oxide	Rb_2O	Rubidium Sulfide	Rb_2S
Cesium Oxide	Cs_2O	Cesium Sulfide	Cs_2S
Francium Oxide	Fr_2O	Francium Sulfide	Fr_2S

Guy poked Professor Terry and whispered, "Isn't water's formula like this?

$$H_2O$$

Professor Terry said, "Good observation, Guy. That's what the formula for water is saying. The subscript 2 under the symbol for Hydrogen does mean that in water there are 2 atoms of Hydrogen, and the symbol O with no subscript means there is only one Oxygen atom in water.

Guy was so excited, "Wow! Now I know what the formula for water means. H_2O means water is made up of 2 atoms of Hydrogen and one atom of Oxygen. Now I know and every time I see the formula for water it will mean so much more to me."

The Alkali Metals and the Nitrogens Become Happy Atom

Chapter7

PERIODS
GROUPS

THE PERIODIC TABLE OF THE ELEMENTS

Period	1A	2A	3B	4B	5B	6B	7B	8B	8B	8B	1B	2B	3A	4A	5A	6A	7A	8A
1	1 H 1.00																	2 He 4.00
2	3 Li 6.94	4 Be 4.01											5 B 10.8	6 C 12.0	7 N 14.0	8 O 16.9	9 F 18.9	10 Ne 20.1
3	11 Na 22.9	12 Mg 24.3											13 Al 26.9	14 Si 28.0	15 P 30.9	16 S 32.0	17 Cl 35.5	18 Ar 39.9
4	19 K 39.1	20 Ca 40.0	21 Sc 44.0	22 Ti 47.9	23 V 50.9	24 Cr 51.9	25 Mn 54.9	26 Fe 56	27 Co 58.9	28 Ni 58.6	29 Cu 63.5	30 Zn 65.3	31 Ga 69.2	32 Ge 72.6	33 As 74.9	34 Se 78.9	35 Br 79.9	36 Kr 83.7
5	37 Rb 85.5	38 Sr 87.6	39 Y 88.9	40 Zr 91.2	41 Nb 92.9	42 Mo 95.9	43 Tc {98}	44 Ru 101	45 Rh 103	46 Pd 106	47 Ag 107	48 Cd 112	49 In 114	50 Sn 119	51 Sb 122	52 Te 126	53 I 126	54 Xe 131
6	55 Cs 132	56 Ba 137.	57 - 71	72 Hf 178	73 Ta 180	74 W 183	75 Re 186	76 Os 190	77 Ir 192	78 Pt 195	79 Au 196	80 Hg 200	81 Tl 204	82 Pb 207	83 Bi 208	84 Po 209	85 At 210	86 Rn 222
7	87 Fr 223	88 Ra 226	89-1 03	104 Rf 267	105 Db 268	106 Sg 271	107 Bh 272	108 Hs 270	109 Mt 278	110 Gs 281	111 Rg 280	112 Cn 285	113 Nh 284	114 Fl 289	115 Mc 288	116 Lv 203	117 Ts 294	118 Og 294

57 La 139	58 Ce 140	59 Pr 141	60 Nd 144	61 Pm 145	62 Sm 150.	63 Eu 151	64 Gd 157	65 Tb 159	66 Dy 163	67 Ho 165	68 Er 167	69 Tm 169	70 Yb 173	71 Lu 175
89 Ac 227	90 Th 232	91 Pa 231	92 U 238.	93 Np 237	94 Pu 234	95 Am 243	96 Cm 247	97 Bk 247	98 Cf 251	99 Es 252	100 Fm 257	101 Md 258	102 No 259	103 Lr 262

Sodium left his family on 6th Street and rushed over to 5th Street. He liked finding new ways of creating Happy Atoms. Even more important, he was interested in adding new compounds to his family, as well as, making Periodic Table Land a happier place to live. Going to 5th Street was a chance to do this.

Having arrived at 5th Street, he looked up Nitrogen who was still happy from making compounds with Boron and Aluminum. He was in his garden whistling a happy tune as he fertilized his tomato plants. Sodium quietly approached him. "Nitrogen," Sodium said, tapping him on the shoulder, "I know you just recently learned how to form compounds with the Boron Family, but I have found a new way to create compounds."

Nitrogen put down his garden tools, and stood up saying, "I'm all ears, Sodium. You must tell me about this new way of becoming a Happy Atom that you've dreamed up. Let's go sit on my patio bench in the shade while we talk."

Sodium told Nitrogen of his adventures on 6th Street that had worked so well. "My Alkali Metals are over on 6th Street, as we speak, forming all sorts of compounds with Oxygen and Sulfur. I told them that I'd send for them, if I managed to form a compound with you. Sodium went on to say, "It's most probable, that together, we will succeed in forming a compound here on 5th Street. You can hope we do, as both of our families will then have a new way to become Happy Atoms. It just makes sense that, if this new method worked on 6th Street, it will work here too."

Nitrogen said "I'd really like you to explain this new method to me."

"Let me tell you how it works. I know you 5th Street elements have five electrons in your Outside Energy Level. I think you remember that becoming a compound requires you to have 8 electrons in your Outside Energy Level. That means, **you need three more electrons to have 8 electron in your Outside Energy Level to be complete.** Look at the model I drew for you and see the 3 empty boxes."

Each of the Alkali Metals on 1st Street has only one electron to give away.

"If we are going to have 3 electrons to offer your elements on 5th Street, we need to gather together in teams of three."

At that point Sodium's team popped out from behind the bush and said, "Here we are, Nitrogen. What do you think of us? Count our electrons. "

Nitrogen said, "By golly, you have three electrons to give away, and I need 3 electrons. I do believe it will work. Let's see if it will really work." At that moment Phosphorus came along. He was told the story, and he saw the possibility that this could work for him too. Both members of the Nitrogen family were convinced that they most likely could form new compounds, and become happy. Phosphorus hung around to watch what would happen.

Sodium said, "We will start as soon as we manage to line up. Nitrogen, you can go first. It's done exactly like the way we formed compounds before. Only this time there will be three Alkali Metal atoms on the left."

Sodium and Nitrogen Form Compounds and Become Happy Atoms

3 **Sodium atoms,** the **metals** that have electrons to give away stand on the left. **Nitrogen**, the **non-metal**, that needs to get the electrons stands on the right.

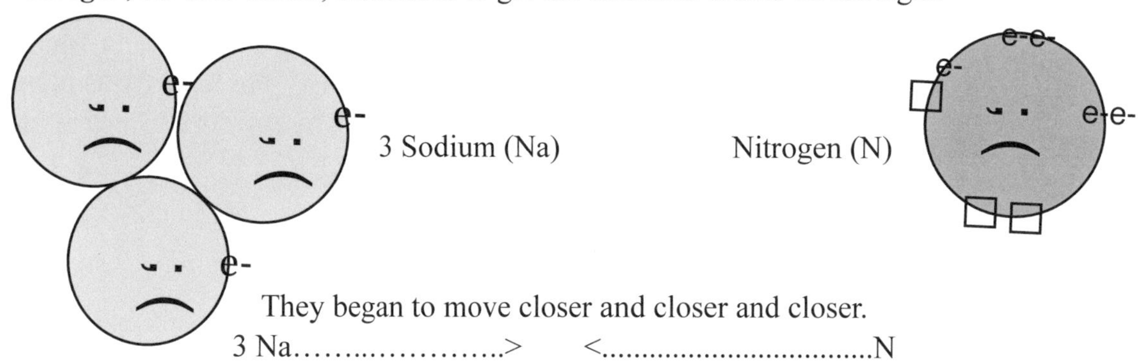

They began to move closer and closer and closer.

3 Na……..…………..> <..................................N

The tension in the air grew. Suddenly, there was a breathtaking sound precisely when Sodium's three electrons jumped into the empty spaces in Nitrogen's Outside Energy Level.

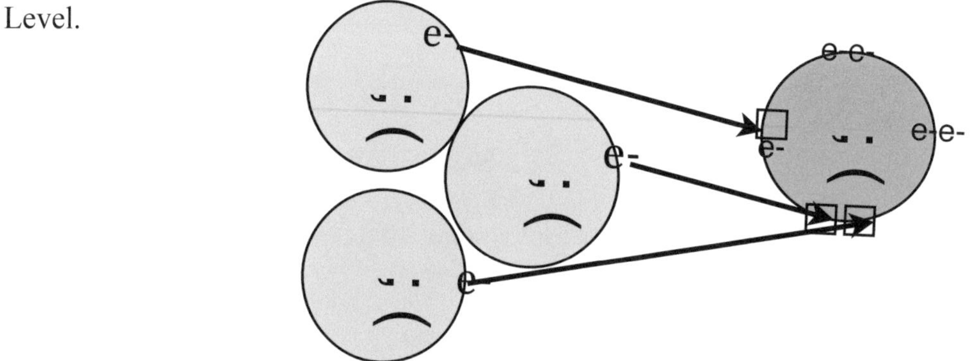

The atoms were joined together to create a new compound. Notice Sodium's 3 electrons in Nitrogen's empty spaces. All 3 Sodium atoms' Complete Energy Levels popped up. They all became Happy Atoms. Look at their smiling happy faces.

A new compound was formed.

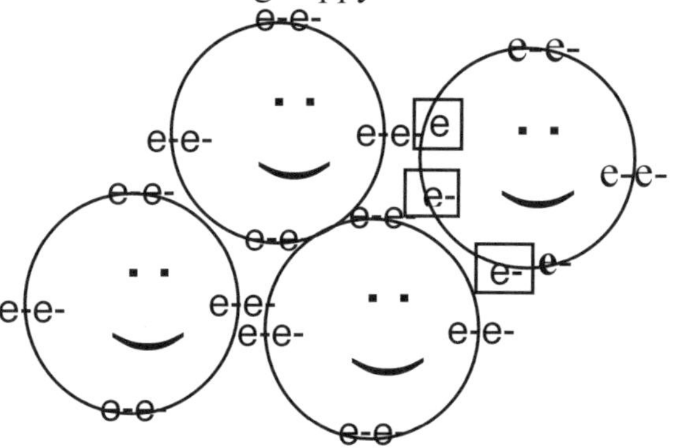

They have a **new name**….. ..Sodium Nitr***ide.***
The symbols became a **formula.** Na_3N

The subscript 3 means that it took 3 atoms of Sodium (Na)
to combine with one atom of Nitrogen (N) to form Na_3N

Notice the non-metal Nitrogen changed the ending of his name.
Nitr***ogen*** dropped ***ogen*** and added ***ide***
Nitr***ogen*** became Nitr***ide***

Here's how we write what has happened in words and diagrams:

3 Sodium atoms plus 1 Nitrogen atom yields Sodium Nitride

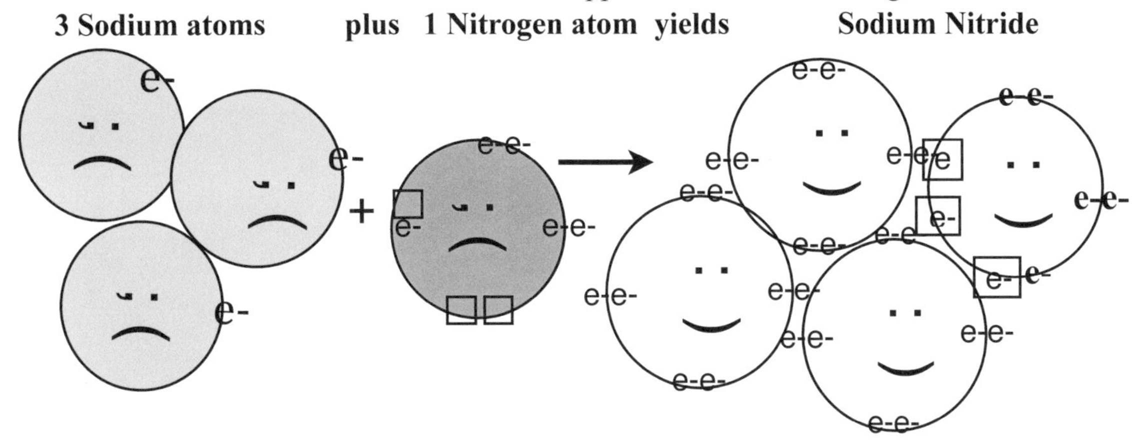

Professor Terry and Guy were watching from the privacy of their bubble. She whispered to Guy, "I must keep reminding you that when we advance to the next level in chemistry, I will teach you how scientists write what happens in symbols and formulas. Now. we're just writing what happens in words and diagrams." Then they returned to watching what was going on with the Alkali Metals and the Nitrogen Family.

****

Phosphorus noticed how Nitrogen was able to form a compound with the Alkali Metal and was anxious to form a compound, too. He wanted to become a Happy Atom in this new way.

Sodium said, "I have a group of three Sodium atoms with me. You can create the new compound with them." Phosphorus looked interested. So Sodium said, "We'll start as soon as we can line up. It's done exactly the way we did it before, only this time there will be 1 Phosphorus atom on the right and 3 Sodium atoms on the Left."

Sodium and Phosphorus Form Compounds and Become Happy Atoms

3 Sodium atoms, the **metal**, that has electrons to give away, stands on the left.
Phosphorus, the **non-metal,** that needs to get the electrons, stands on the right.

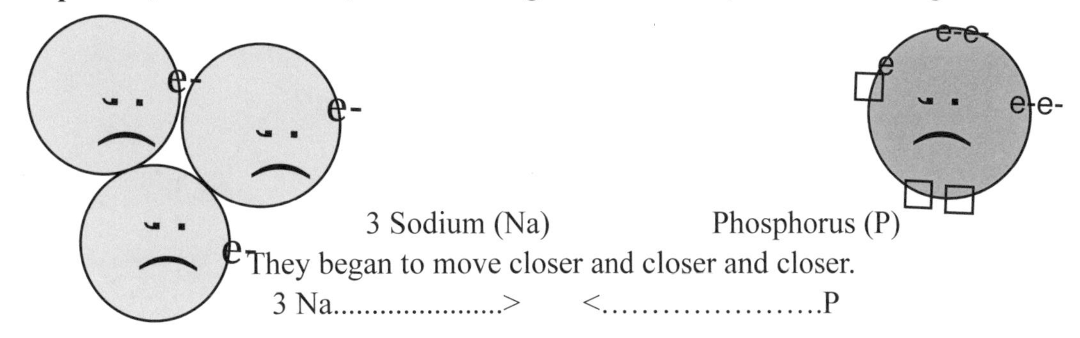

3 Sodium (Na) Phosphorus (P)

They began to move closer and closer and closer.

3 Na.....................> <......................P

"Watch how the whole process evolved. Notice especially the electrons. Each Sodium atom aimed his electron at an empty space in the Phosphorus atom and then released the electron to fly over to Potassium's atom.

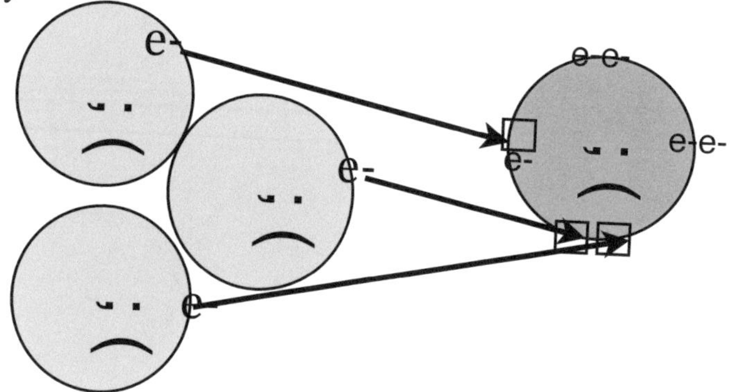

The tension in the air grew. Suddenly, a breathtaking sound occurred precisely when Sodium's three electrons jumped into the 3 empty spaces in Phosphorus' Outside Energy Level. Again it was magical. Phosphorus got a complete Outside Energy Level. The 3 Sodium atoms' complete energy levels, hidden away, popped up when they gave away the 1 electron. Then they all had complete Outside Energy Levels. They became Happy Atoms. The three Sodium atoms and one Phosphorus atom were joined together to create a new compound.

See how happy they are because
they all have complete Outside Energy Levels.

And there's more!

They get a **new name.**... ..Sodium Phosph***ide***
The symbols become a **formula.** Na_3P

The subscript 3 under the Na shows that it took 3 atoms of Sodium
to combine with 1 Phosphorus atom.
Notice Phosphorus the non-metal dropped its ending ***orus***
*Phosph**orus** became Phosph**ide***

Here's what happened written in words and diagrams
3 Sodium atoms plus 1 Phosphorus atom yields. Sodium Phosph*ide*

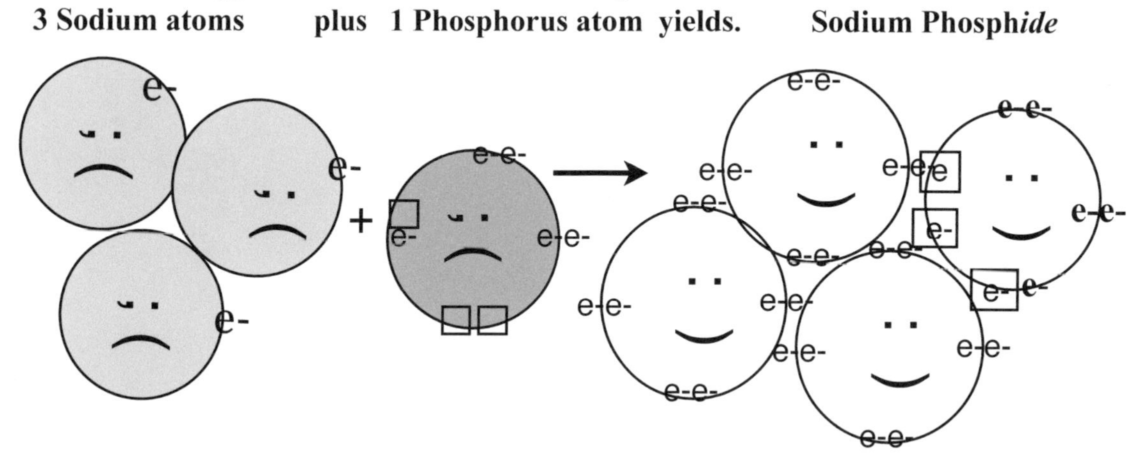

All Alkali Metals Form Compounds with Nitrogen and Phosphorus

Nitrogen and Phosphorus were delighted that they had learned another way to create new compounds and become even happier Happy Atoms. Sodium was delighted also.

Sodium said, "Notice that both of you formed the compounds in exactly the same way. There are five more Alkali Metals: Lithium, Potassium, Rubidium, Cesium, and Radon. Both Nitrogen and Phosphorus can form compounds with each of these Alkali Metals in exactly the same way. That means you each can make 5 more compounds with those Alkali Metals I just named. Along with the two compounds you made with me, that will make a total of 12 new compounds to bring more happiness to 5th Street and your Nitrogen Family."

Nitrogen and Phosphorus said, "Go get the rest of your Alkali Metals. Forming compounds will not only bring happiness to our Nitrogen family, but also it will be good for your family and the whole of Periodic Table Land. It wasn't too long ago when everyone was sad. Life is good now."

Sodium went back to 6th Street where the Alkali Metals were waiting to hear from him. He told them of his success in forming compounds on 5th Street. Then he told them to get ready saying,"Gather together three atoms just like your own and come to 5th Street with me to form compounds with Nitrogen and Phosphorus." So they all gathered in teams of three, and jumped into the bus-size bubble below the family balloons. With a twirl of the magic wand, the balloons lifted them high in the sky and kept on course to 5th Street. Shortly it brought the Alkali Metals to a soft landing near the home of Nitrogen and Phosphorus.

Sodium led all the teams of the Alkali Metals to meet Nitrogen and Phosphorus. Nitrogen had a plan already mapped out. The new compounds were formed in an orderly way exactly as they had done before. Groups of three atoms of the Alkali Metals aimed their electrons toward the 3 empty spaces in the Outside Energy Levels of Nitrogen and Phosphorus and so many new compounds were formed.

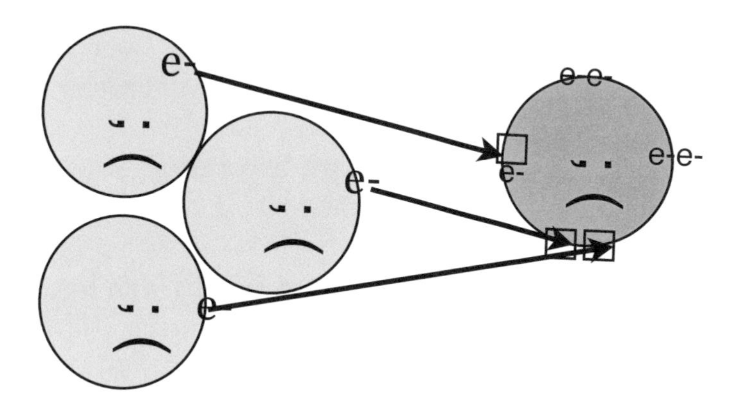

Sodium showed his family how all the new compounds were formed. They were really excited. Then Sodium showed how this was written in words and diagrams.

Here's the diagram that shows exactly how each of the compounds was formed.

3 Alkali Metal Atoms + 1 Nitrogen Family Atom yields Compound

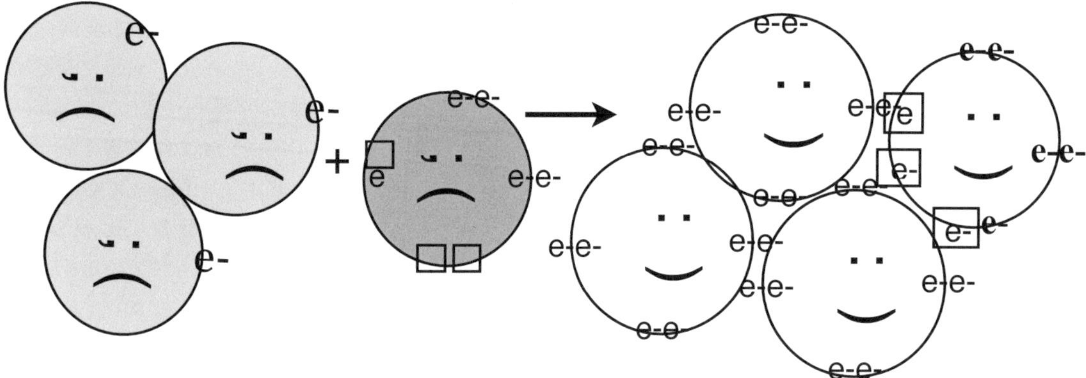

Here are all the compounds that were formed with their new name and formula.

Alkali Metal + Nitrogen

NAME	FORMULA
Lithium Nitride	Li_3N
Sodium Nitride	Na_3N
Rubidium Nitride	Rb_3N
Potassium Nitride	K_3N
Cesium Nitride	Cs_3N
Francium Nitride	Fr_3N

Alkali Metal + Phosphorus

NAME	FORMULA
Lithium Phosphide	Li_3P
Sodium Phosphide	Na_3P
Rubidium Phosphide	Rb_3P
Potassium Phosphide	K_3P
Cesium Phosphide	Cs_3P
Francium Phosphide	Fr_3P

"Guy instead of a Quiz, tell me what you know about The Alkali Metals forming a compound with Nitrogen Family."

Guy listed the steps he remembered.

1. All the elements in Group 1A have 1 electron in their Outside Energy Level.
2. All the elements in Group 5A have 5 electrons in their Outside Energy Level.
3. To be Happy Atoms Group 1A elements need to give away their 1 electron to have their complete energy level pop up and become their complete Outside Energy Level.
4. Group 5A elements need to take in 3 electrons to make their Outside Energy Level complete. They have 5 electrons and need 8 to be complete.
5. For these 2 families to form a compound and become a Happy Atom it takes 3 Alkali Metal elements to give the Nitrogen family elements the 3 electrons they need to become a new compound and a Happy Atom.
6. The formula for Sodium to combine with Nitrogen is Na_3N. The compound's name is Sodium Nitr*ide*. The non-metal changed his name's ending to *ide.* Nitrogen changed to Nitr*ide*. Phosphorus would change to Phosph*ide.*

Professor Terry said, "Good job, Guy. You got the idea. Now we can move on to our next adventure."

Alkaline Earth Metals and the Halogens Become Happy Atoms
Chapter 7

THE PERIODIC TABLE OF THE ELEMENTS

PERIODS ↓ GROUPS →

	1A	2A	3B	4B	5B	6B	7B	8B			1B	2B	3A	4A	5A	6A	7A	8A
1.	1 H 1.00																	2 He 4.00
2.	3 Li 6.94	4 Be 4.01											5 B 10.8	6 C 12.0	7 N 14.0	8 O 16.9	9 F 18.9	10 Ne 20.1
3.	11 Na 22.9	12 Mg 24.3											13 Al 26.9	14 Si 28.0	15 P 30.9	16 S 32.0	17 Cl 35.5	18 Ar 39.9
4.	19 K 39.1	20 Ca 40.0	21 Sc 44.0	22 Ti 47.9	23 V 50.9	24 Cr 51.9	25 Mn 54.9	26 Fe 56	27 Co 58.9	28 Ni 58.6	29 Cu 63.5	30 Zn 65.3	31 Ga 69.2	32 Ge 72.6	33 As 74.9	34 Se 78.9	35 Br 79.9	36 Kr 83.7
5.	37 Rb 85.5	38 Sr 87.6	39 Y 88.9	40 Zr 91.2	41 Nb 92.9	42 Mo 95.9	43 Tc {98}	44 Ru 101	45 Rh 103	46 Pd 106	47 Ag 107	48 Cd 112	49 In 114	50 Sn 119	51 Sb 122	52 Te 126	53 I 126	54 Xe 131
6.	55 Cs 132	56 Ba 137.	57 - 71	72 Hf 178	73 Ta 180	74 W 183	75 Re 186	76 Os 190	77 Ir 192	78 Pt 195	79 Au 196	80 Hg 200	81 Tl 204	82 Pb 207	83 Bi 208	84 Po 209	85 At 210	86 Rn 222
7.	87 Fr 223	88 Ra 226	89-103	104 Rf 267	105 Db 268	106 Sg 271	107 Bh 272	108 Hs 270	109 Mt 278	110 Gs 281	111 Rg 280	112 Cn 285	113 Nh 284	114 Fl 289	115 Mc 288	116 Lv 203	117 Ts 294	118 Og 294

57 La 139	58 Ce 140	59 Pr 141	60 Nd 144	61 Pm 145	62 Sm 150.	63 Eu 151	64 Gd 157	65 Tb 159	66 Dy 163	67 Ho 165	68 Er 167	69 Tm 169	70 Yb 173	71 Lu 175
89 Ac 227	90 Th 232	91 Pa 231	92 U 238.	93 Np 237	94 Pu 234	95 Am 243	96 Cm 247	97 Bk 247	98 Cf 251	99 Es 252	100 Fm 257	101 Md 258	102 No 259	103 Lr 262

Sodium had a new mission. He slipped into his bubble, with the beautiful light green balloons and set his course toward 2nd Street. Finally, he reached his destination, the home of the Alkaline Earth Metals. He walked down 2nd Street to #20 where his old buddy Calcium lived. He stuck his head into the open window and called, “I’ve got some good news.”

Calcium said, “Come on in, Sodium. Sit down and we can have a nice visit.”

Bursting with enthusiasm to share the good news with his friend, Sodium couldn’t even wait to sit down and so began.“There’s a new way of making Happy Atoms, and I can’t wait another second to share this with you.”

Calcium had learned that there was only one way to form a compound. That way was to join with elements that needed exactly the same number of electrons that he had to give away. Calcium found it difficult to believe that there was any other way to become a Happy Atom. Calcium said, “I have 2 electrons to give away, and there is only one family that needs just 2 electrons in all of Periodic Table Land—the Oxygen Family.”

Calcium was more than a little skeptical about what Sodium was saying. "How can you say I could go to any other street and and form compounds. That's ridiculous. The whole idea goes against everything I know. I need to make compounds with elements that need to get exactly 2 electrons."

Sodium tried to calm him down."Calcium, you need to listen. If you are open minded, you will see that what I say is easy to understand. Just listen to me. You have 2 electrons to give away. The Halogens, on 7th Street, can only take one electron. So you tell them to go get another Halogen atom to take your other electron. He does, and you get rid of both your electrons, and two Halogen atoms get complete Outside Energy Levels. So one Alkaline Earth Metal atom can join up with 2 Halogen atoms and form a compound. That's all there is to it. The whole process proceeds as before. The big secret to make this work is for the Halogens to always come in pairs to be able to take both of your electrons. Two Halogens get complete Outside Energy Levels. You'll see. Compounds definitely can be formed this way. I've seen it done. I have a diagram to show you how this works. Look at this diagram."

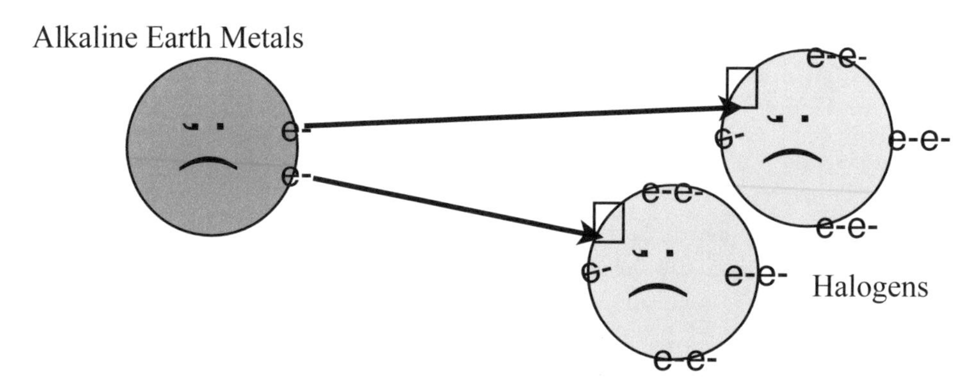

Calcium said, "The diagram helped me understand what you have been saying. I can see the 2 Halogen atoms can easily take my 2 electrons. It's logical, and you saw it happen. I guess I'm willing to try it."

"You will be happy that you made this decision," said Sodium, pleased that Calcium understood how it could happen. "The process is the same as always. You line up as usual; move closer and closer as you did before, and your electrons will jump into the empty spaces in both Halogen atoms. You all will become Happy Atoms. I've done something just like it. It works. I'm so sure it will work, I encourage you to take along all the elements in your family right away and go to 7th Street immediately. You will all become Happy Atoms, and form lots of new compounds."

Finally convinced that this new method would work, Calcium sent for his whole family. Beryllium, Magnesium Strontium Barium and Radium came curious as to why Calcium had sent for them.

When they arrived Calcium said "Gather round and listened carefully I've discovered something exciting. There's a new way to become a Happy Atom and form lots of new compounds." Calcium taped the diagram on the wall and said, "Look at this diagram."

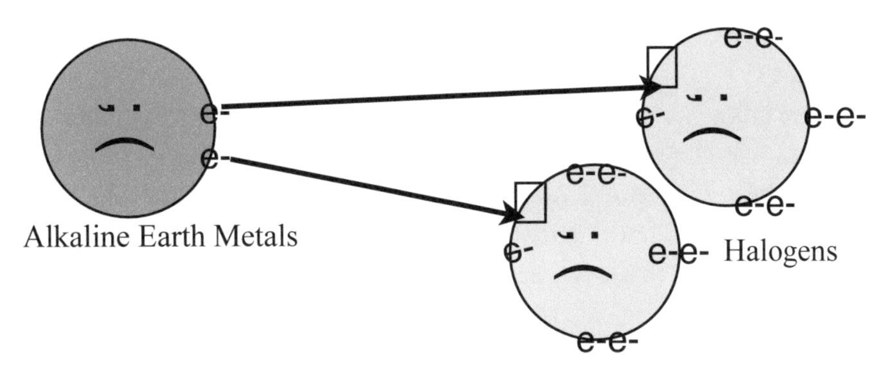

Pointing to the diagram Calcium began,"As you know, we, the Alkaline Earth Metals, have two electrons in our Outside Energy Level. Again pointing to the diagram, he said, "That's what our Outside Energy Level looks like. The Halogens can only take in one of our electrons. So we need to tell the Halogens to get another atom to take our second electron. The arrows show how the electrons will move into the empty space in each of the 2 Halogen atoms. It's simple. So, let's get going and see the Halogens. It's time to start making more Happy Atoms."

Calcium got out the magic wand, and with several waves of the wand the size of the bubble adjusted to accommodate the entire family. The Alkaline Earth Metals piled into the bubble, fastened their seat belts, and with a twirl of the magic wand the lovely amber colored balloons lifted the family bubble. There wasn't a cloud in sight. It was a beautiful day, and the flight was smooth. On the way Calcium explained what to expect."When we get to 7th Street, I will meet with Chlorine, and you can watch me form a compound in this new way. Then you'll be free to meet up with any of the Halogens you wish and form compounds with them. Just tell them to come in pairs."

Soon they arrived and landed in the clearing at the top of 7th Street. Calcium got out first and made his way down the path through the pine forest kicking a few pine cones on the way. The others followed and stayed at the edge of the pine forest. Calcium continued on. He passed by Fluorine's unique tooth shaped house on his way to locate Chlorine. Ahead was #17, 7th Street, Chlorine's lovely home. The washing machine was going at Chlorine's place, and there was a refreshing smell of bleach coming out of the laundry room. Chlorine came out to meet Calcium who said directly, "I've learned a new way to make compounds. I've heard this is a great way to become a Happy Atom."

Chlorine said, "I'm ready if you are. Just give me one of your electrons, Calcium. Becoming a Happy Atom is seconds away. I feel it already."

"Not so fast Chlorine. Maybe you didn't notice I have 2 electrons. If I give you one electron, you will be happy all right; but I will still be stuck with one electron in my Outside Energy Level, and I'll be very sad. As you know when that energy level is not complete, I will not be happy. Let me explain what you have to do. You have to get another Chlorine atom to take my other electron. Then we will all be Happy Atoms together." Chlorine understood. He went off to find another Chlorine atom.

Back at the edge of the forest, the Alkaline Earth Metals watched the action as they peeked between the branches of the fragrant pine trees. Two of them climbed up a sturdy tree and had a grand view. Professor Terry and Guy were close to the action, invisible as they sat inside their magic bubble. Guy was learning compound formation from a front row seat.

Two Chlorine atoms soon returned and were ready to make a new compound with Calcium. The routine was the same as usual. By this time everyone knew the routine by heart. Metal on the left. Non-metal on the right. Start moving. But this time it was different so they all looked on carefully so see how the compound was formed this new way. They were ready to learn this new method.

Calcium and Chlorine Form Compounds and Become Happy Atoms

Calcium, **a metal**, the atom with electrons to give away, stood on the left.
Chlorine's 2 atoms, **non-metals**, the atoms that needed electrons, stood on the right.

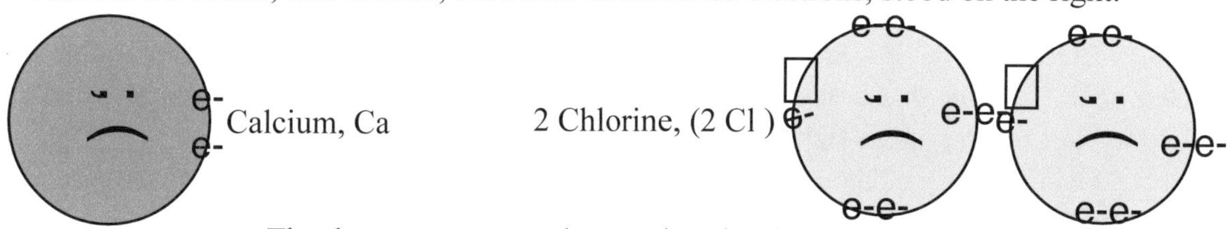

The three atoms moved toward each other.
1 Ca ....>   ……..,,,,2 Cl

Calcium's 2 electrons aimed at Chlorine's empty spaces where they needed to go.

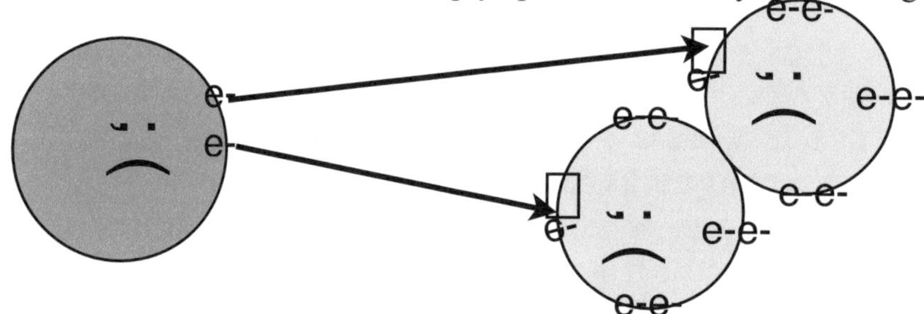

When they were close enough, the 2 Calcium electrons jumped into the empty space in each of the 2 Chlorine atoms. Calcium's complete energy level popped up.That famous sound rang out the joy of that moment.
Calcium and Chlorine became Happy Atoms.
Look at the big smiles on their Happy Atom faces!
They were truly happy. But there's more.
They are something very new.

Calcium and Chlorine became a compound.

*They received a **new name**...............Calcium Chlor**ide***

*Their symbols changed into a **formula** .....Ca**Cl**$_2$*

*Notice: You must put subscript 2 under the symbol for Cl. Because it took 2 atoms of Chlorine to make this compound. For Chlorine we wrote **Cl**$_2$.*

Again the non-metal changed the ending of his name.

*Chlor**ine** became Chlor**ide***

This is how to write what happened in words and diagrams:

Calcium plus 2 Chlorine atoms yields Calcium Chloride

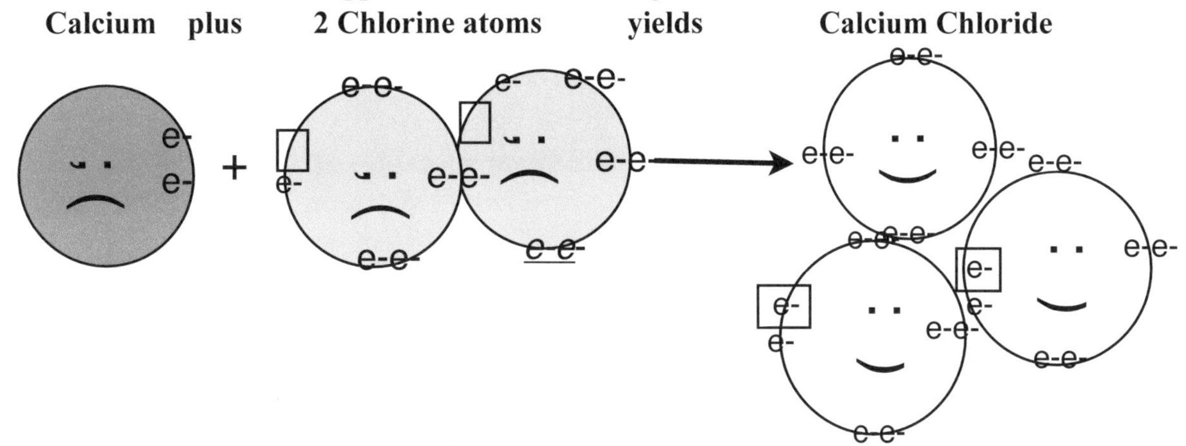

Calcium said, “I have my whole family with me. I suggest you get the rest of the Halogens, and we will have two very happy families.”

Chlorine went from house to house on 7th Street spreading the good news. All the Halogens. Fluorine, Bromine, Iodine and Astatine all gathered together in the grassy field. They all were so pleased to think there was a new way to make compounds. They never believed they would be able to make compounds with the Alkaline Earth Metals.

Calcium explained the process to the eager Halogens. He drew a diagram to show how it worked. “Halogens, come look at this diagram. The secret is to come in pairs. The arrows show how the Alkaline Earth Metal’s 2 electrons will move into each of your empty spaces.”

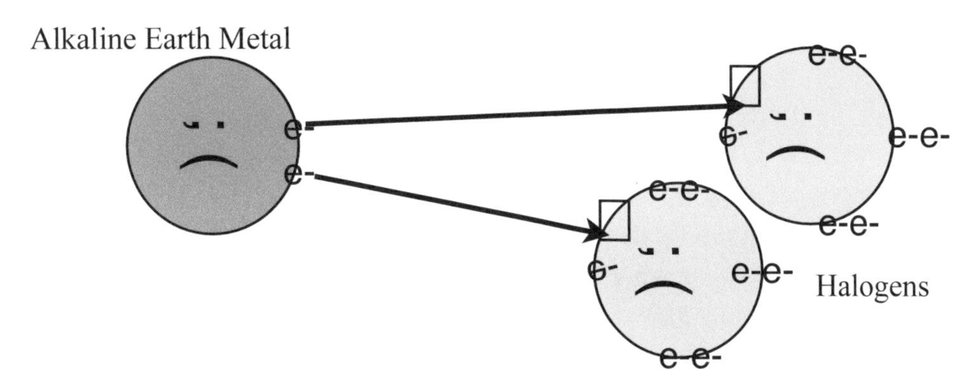

“The routine to make this happen is the same as before. The only difference is that on the right you must have 2 Halogens. You line up and then move closer and closer. When close enough, the electrons jump and you become Happy Atoms. A new compound is formed. In the formula you have to write the number of atoms that are in the compound. So the Halogen writes a 2 as a subscript beneath the Halogen symbol."

Calcium ever the teacher said, “Here’s a model for you to follow. Each of you Halogens: Fluorine, Chlorine, Bromine, Iodine, and Astatine can make a compound with each of my Alkaline Earth Metals: Beryllium, Calcium, Magnesium, Strontium, Barium, and Radon. Follow this example and you will all become Happy Atoms in this new exciting way.

All The Alkaline Earth Metals and the Halogens Form Compounds and Become Happy Atoms

The Alkaline Earth Metal, the **metal** with electrons to give away, is on the **left.**
The 2 **Halogen** atoms, the **non-metals** that need to get electrons, lined up on the right.

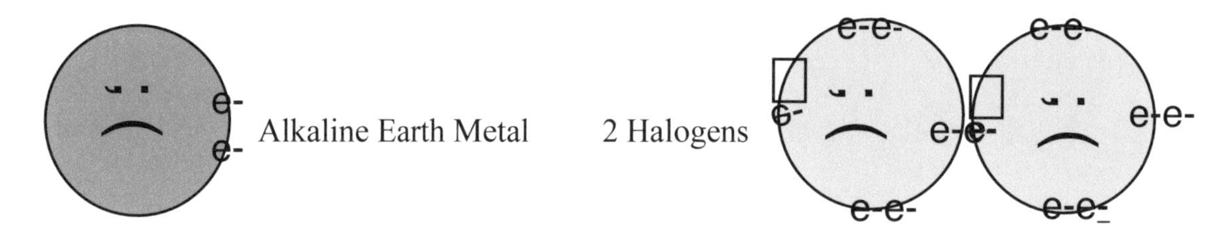

Then the Alkaline Earth Metal and Halogens moved closer and closer
Alkaline Earth Metal ....> <......................2 Halogen atoms
The Alkaline Earth Metal's electrons aim at the empty spaces in the Halogens

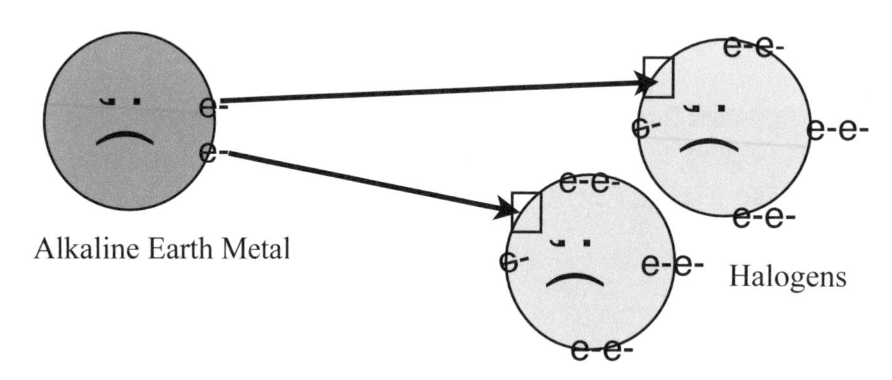

The arrows show the way the electrons move toward the empty spaces in the Halogens.

Then they meet, and the electrons jump from the Alkaline Earth Metal into the empty spaces in the Halogen's two atoms. Suddenly, there's that now famous, joyful sound celebrating that moment when a new compound is formed.

They get a **new name.** *The ending of the non-metal changes from* ***ine*** *to* ***ide.***

Their symbols change into a **formula**. *Remember to put 2 as a subscript under the symbol for the Halogens. It always takes 2 Halogens to form a compound with 1 Alkaline Earth Metal.*

Alkaline Earth Metals Form Compounds with Halogens

Calcium said, "Well, get busy and make lots of compounds following the model you just saw."

And they did!

Beryllium Calcium, Magnesium Strontium Barium and Radium remembered to tell the Halogens to come in pairs to make these compounds and the Halogens came in pairs. It was very successful and lots of Happy Atoms were formed.

Here Are The Compounds That They Made.

Fluoride Compounds		**Chloride Compounds.**		**Bromide Compounds**	
Beryllium Fluoride	BeF_2	Beryllium Chloride	$BeCl_2$	Beryllium Bromide	$BeBr_2$
Calcium Fluoride	CaF_2	Calcium Chloride	$CaCl_2$	Calcium Bromide	CaBr2
Magnesium Fluoride	MgF_2.	Magnesium Chloride	$MgCl_2$	Magnesium Bromide	$MgBr_2$
Strontium Fluoride	SrF_2	Strontium Chloride	$SrCl_2$	Strontium Bromide	$SrBr_2$
Barium Fluoride	BaF_2	Barium Chloride	$BaCl_2$	Barium Bromide	$BaBr_2$
Radium Fluoride	RaF_2	Radium Chloride	$RaCl_2$	Radium Bromide	$RaBr_2$

And they all became Happy Atoms.

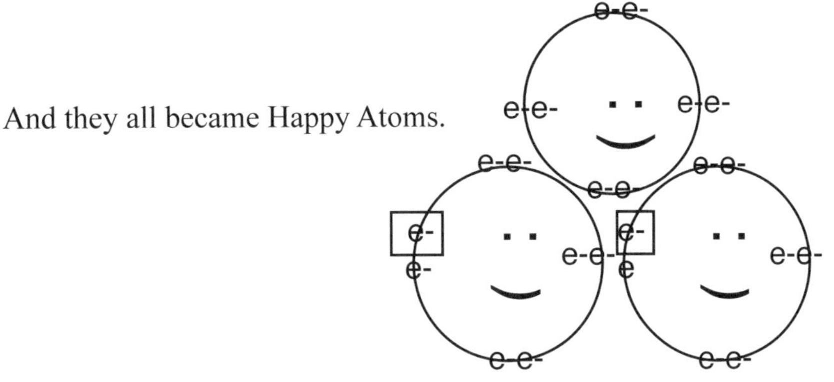

Aluminum hidden by the bushes, observed this new way of forming compounds. He thought, "I bet my family could do this too." Off he went to to his family's home on 3rd Street.

Professor Terry said, "Guy it's quiz time. See if you can write the formulas for the following Alkaline Earth Metals that formed a compound with a Halogens."

QUIZ

1. Strontium Iodide __________________. 2. Beryllium Bromide__________________

Looking at the Periodic Table (Pg 76), Guy said, "Strontium and Beryllium are in Group 2A. That makes them Alkaline Earth Metals with 2 electrons to give away. Iodine and Bromine are Halogens in Group 7A. That means that Iodine and Bromine have 7 electrons and needed only one more electron to be complete. That means it will take 2 Halogens to take the Alkaline Earth Metals' 2 electrons."

Guy was very pleased to announce, "The formulas are:Strontium Iodide——SrI_2
Beryllium Bromide——$BeBr_2$.

Guy showed that he had learned how these Chemical Families became compounds. Along with becoming compounds they became Happy Atoms.

Professor Terry confirmed the answers: SrI_2 and $BeBr_2$

The Boron Family and the Halogens Become Happy Atoms

Chapter 8

THE PERIODIC TABLE OF THE ELEMENTS

PERIODS

GROUPS

Periods	1A	2A	3B	4B	5B	6B	7B	8B	8B	8B	1B	2B	3A	4A	5A	6A	7A	8A
1.	1 H 1.00																	2 He 4.00
2.	3 Li 6.94	4 Be 4.01											5 B 10.8	6 C 12.0	7 N 14.0	8 O 16.9	9 F 18.9	10 Ne 20.1
3.	11 Na 22.9	12 Mg 24.3											13 Al 26.9	14 Si 28.0	15 P 30.9	16 S 32.0	17 Cl 35.5	18 Ar 39.9
4.	19 K 39.1	20 Ca 40.0	21 Sc 44.0	22 Ti 47.9	23 V 50.9	24 Cr 51.9	25 Mn 54.9	26 Fe 56	27 Co 58.9	28 Ni 58.6	29 Cu 63.5	30 Zn 65.3	31 Ga 69.2	32 Ge 72.6	33 As 74.9	34 Se 78.9	35 Br 79.9	36 Kr 83.7
5.	37 Rb 85.5	38 Sr 87.6	39 Y 88.9	40 Zr 91.2	41 Nb 92.9	42 Mo 95.9	43 Tc {98}	44 Ru 101	45 Rh 103	46 Pd 106	47 Ag 107	48 Cd 112	49 In 114	50 Sn 119	51 Sb 122	52 Te 126	53 I 126	54 Xe 131
6.	55 Cs 132	56 Ba 137.	57 - 71	72 Hf 178	73 Ta 180	74 W 183	75 Re 186	76 Os 190	77 Ir 192	78 Pt 195	79 Au 196	80 Hg 200	81 Tl 204	82 Pb 207	83 Bi 208	84 Po 209	85 At 210	86 Rn 222
7.	87 Fr 223	88 Ra 226	89-1 03	104 Rf 267	105 Db 268	106 Sg 271	107 Bh 272	108 Hs 270	109 Mt 278	110 Gs 281	111 Rg 280	112 Cn 285	113 Nh 284	114 Fl 289	115 Mc 288	116 Lv 203	117 Ts 294	118 Og 294

57 La 139	58 Ce 140	59 Pr 141	60 Nd 144	61 Pm 145	62 Sm 150.	63 Eu 151	64 Gd 157	65 Tb 159	66 Dy 163	67 Ho 165	68 Er 167	69 Tm 169	70 Yb 173	71 Lu 175
89 Ac 227	90 Th 232	91 Pa 231	92 U 238.	93 Np 237	94 Pu 234	95 Am 243	96 Cm 247	97 Bk 247	98 Cf 251	99 Es 252	100 Fm 257	101 Md 258	102 No 259	103 Lr 262

Aluminum was up on the flag pole untangling the Boron Family flag when Sodium arrived to talk to him about the new way to form compounds. Aluminum slid down the pole to greet Sodium who couldn't wait to share the good news. "I came here to suggest that your family could create compounds with the Halogens on 7th Street."

Aluminum a little embarrased said, "I hope you don't think I'm a snoop, but I just happened to be over by the Halogens where I watched the Alkaline Earth Metals creating compounds with them. They asked the Halogens to get two of their atoms to combine with theirs because they had 2 electrons to give away. So I know it can be done. It made me think that since my family had 3 electrons to give away, we could ask the Halogens to get 3 of their atoms and we could also form a compound. I was planning to ask you if there were any details that I needed to know about."

Sodium assured Aluminum that it was straight forward. You have 3 electrons to get rid of. The Halogens can only take 1 electron. So you have to tell the Halogens to come in groups of three to take all 3 of your electrons. If they do, you both will become

Happy Atoms. As you observed, the routine is the same as it always was for making compounds. Did you learn how to write chemical formulas for compounds made this way?

"Yes," said Aluminum."I know the formula has to tell how many of each element. joined to make that compound. To show that, we write subscripts."

Sodium wished him good luck, and returned home to 1st Street.

Aluminum left deciding not to share this good news with Boron until he actually became a Happy Atom in this new way. If it was a good plan, he would come back to get Boron. Aluminum was ready to create a new compound on 7th Street and become a Happy Atom. His shiny silver balloons landed in the clearing in the pine forest behind 7th Street. Then Aluminum made his way past Fluorine and Chlorine's houses to the home of Bromine.

Aluminum found Bromine sitting out on the side porch of his house. He joined him and shared the information Sodium had told him. "If you gather together 3 Bromine atoms, each of your atoms could take one of my three electrons. You would have 3 atoms with complete Outside Energy Levels and they will become Happy Atoms. I would get rid of my 3 electrons, and my complete energy level should pop up, and I'll be a Happy Atom, too. We'll all be Happy Atoms, and together we'll form a new compound. I'm still not sure it will work, but Sodium was convinced it will."

Bromine said, "It definitely will work. I can only take one electron; but when the Alkaline Earth Metals were here with two electrons to give away, we created new compounds together. I came with 2 Bromine atoms, and we each took one electron from Magnesium, who was able to get rid of both his electrons. We became a new compound, Magnesium Bromide. Now, to form a compound with you, I have to find three Bromine atoms, and we will definitely become Happy Atoms."

Aluminum said, "That's what Sodium said would happen! I guess he was right. I'll relax here on your porch while you round up 3 Bromine atoms, and together we'll create a new compound."

Bromine had a hard time getting his two friends to come back with him. They were screaming and complaining that this was a waste of time. Bromine said, "Where were you when the Alkaline Earth Metals were here? We made beautiful new compounds with them. It really works."

They finally came back with Bromine, but they were not happy about coming. They said, "We're happy making compounds the simple way with the elements that have only one electron to give away. Why did you get such a crazy new idea?" Bromine convinced them to be polite and give it a try. At last, Bromine returned to Aluminum who was waiting patiently for Bromine to arrive with three Bromine atoms.

Aluminum could see that they had no idea about how this new method worked. So he said, "I have a diagram on this paper showing how the process works. Come over and look. It will show you how my electrons will help you three Halogen atoms achieve a complete Outside Energy Levels."

The three Bromine atoms gathered round Aluminum. Aluminum placed the diagram flat down on the low table in front of them. They bent over the table, looked closely at the diagram, and saw immediately how Aluminum's electrons would make their atoms complete.

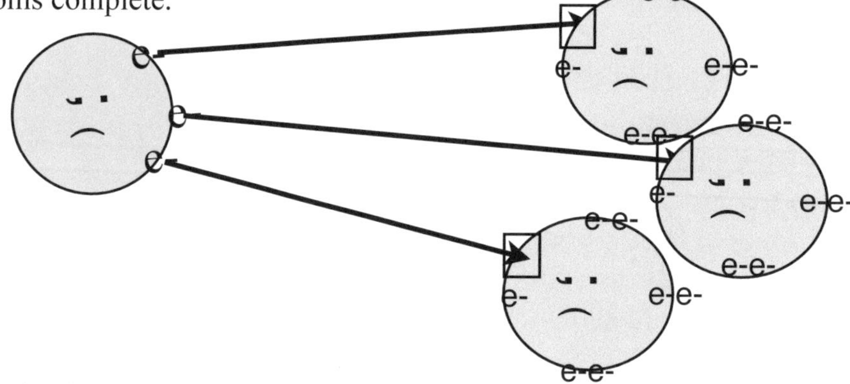

Aluminum told the Bromine atoms that the ritual is much the same as it always was. The only difference is that on the right we have 3 of your Bromine atoms instead of one atom. They lined up properly.

Aluminum and Bromine Become Happy Atoms

Aluminum, the **metal,** who had electrons to give away lined up on th**e** left.
The 3 **Bromine** atoms, the **non-metals** that needed to get the electrons stood on the right**.**

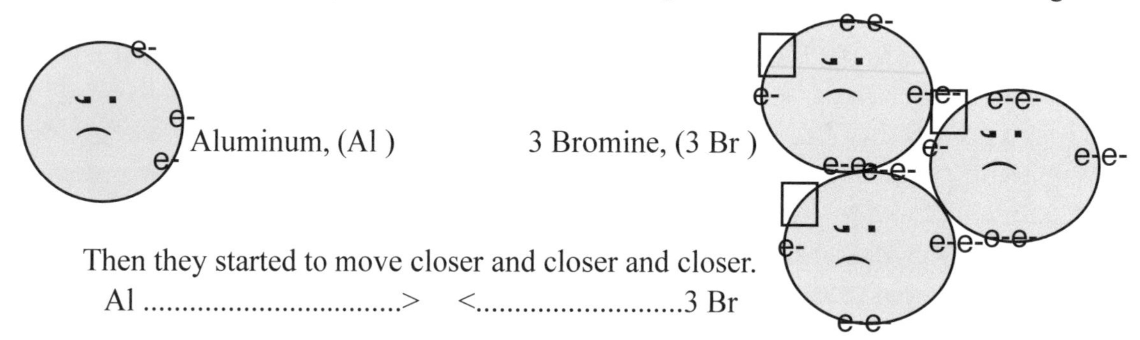

Then they started to move closer and closer and closer.
Al ....> <..........................3 Br

The arrows show how Aluminum's electrons aim at the empty space in each of the three Bromine atoms.

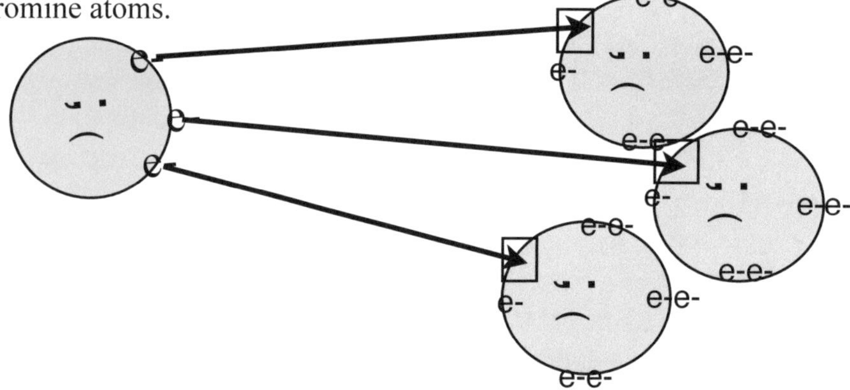

At that magical moment, Aluminum's 3 electrons moved into Bromine's empty spaces and they all became Happy Atoms. The joyful fanfare happened celebrating the moment when they all became a new creation, a compound. They were indeed Happy Atoms. Look at all their smiling faces. Aluminum got rid of all 3 of his electrons and his complete Outside Energy Level popped up. Each of the Bromine atoms got an electron making all 3 of their Outside Energy Levels complete. They had every reason to be Happy Atoms.

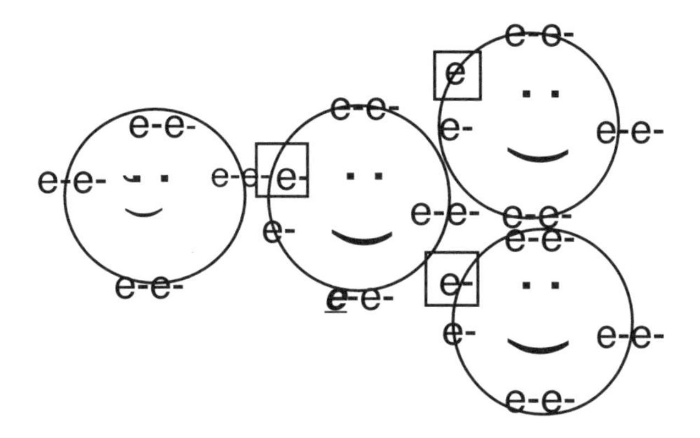

Not only did they become Happy Atoms, they became a new creation, a compound with a new name and a formula.

The compound has a new name........................Aluminum Brom**ide.**
Their symbols changed to a formula……………………..$Al\mathbf{Br}_3$
The subscript 3 under the symbol, Br_3 means 3 Bromine atoms joined with 1 Aluminum atom to create this compound, $Al\mathbf{Br}_3$. Check it out. Counting, there were 3 Bromine atoms joined to 1 Aluminum atom.

I hope that you noticed Brom***ine*** changed his name to Brom***ide.*** The ***ine*** changed to ***ide.*** Whenever 2 elements join together to form a compound, the non-metal changes the ending of his name to ***ide***.

Here's how Aluminum and Bromine became Aluminum Brom**ide** in words and diagrams:
Aluminum plus 3 Bromine atoms yields Aluminum Bromide

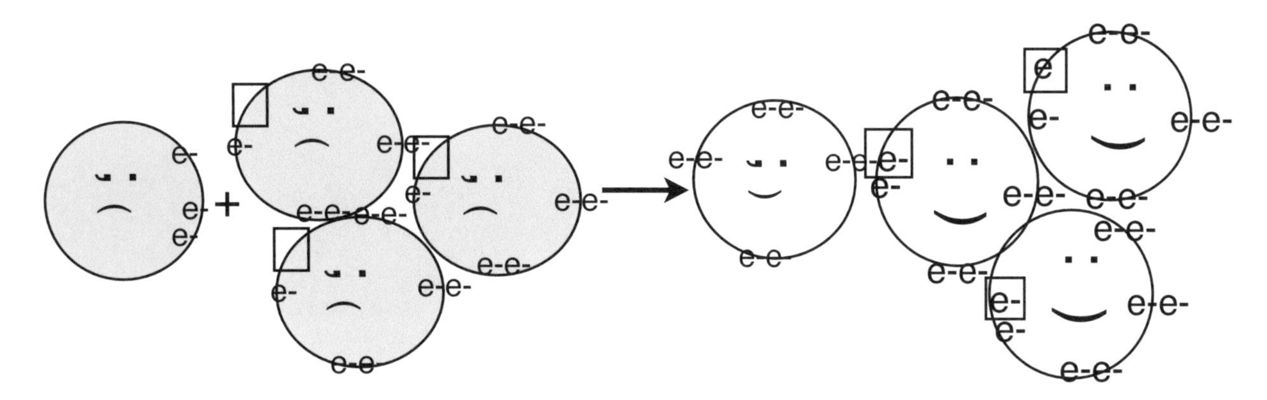

Aluminum was really excited. He now knew that this new way of creating compounds worked. So he flew back to 3rd Street to tell Boron.

"Boron, here are the rules. This is how I became a new compound, a Happy Atom." Aluminum hung the RULES on the wall and began explaining them to Boron.

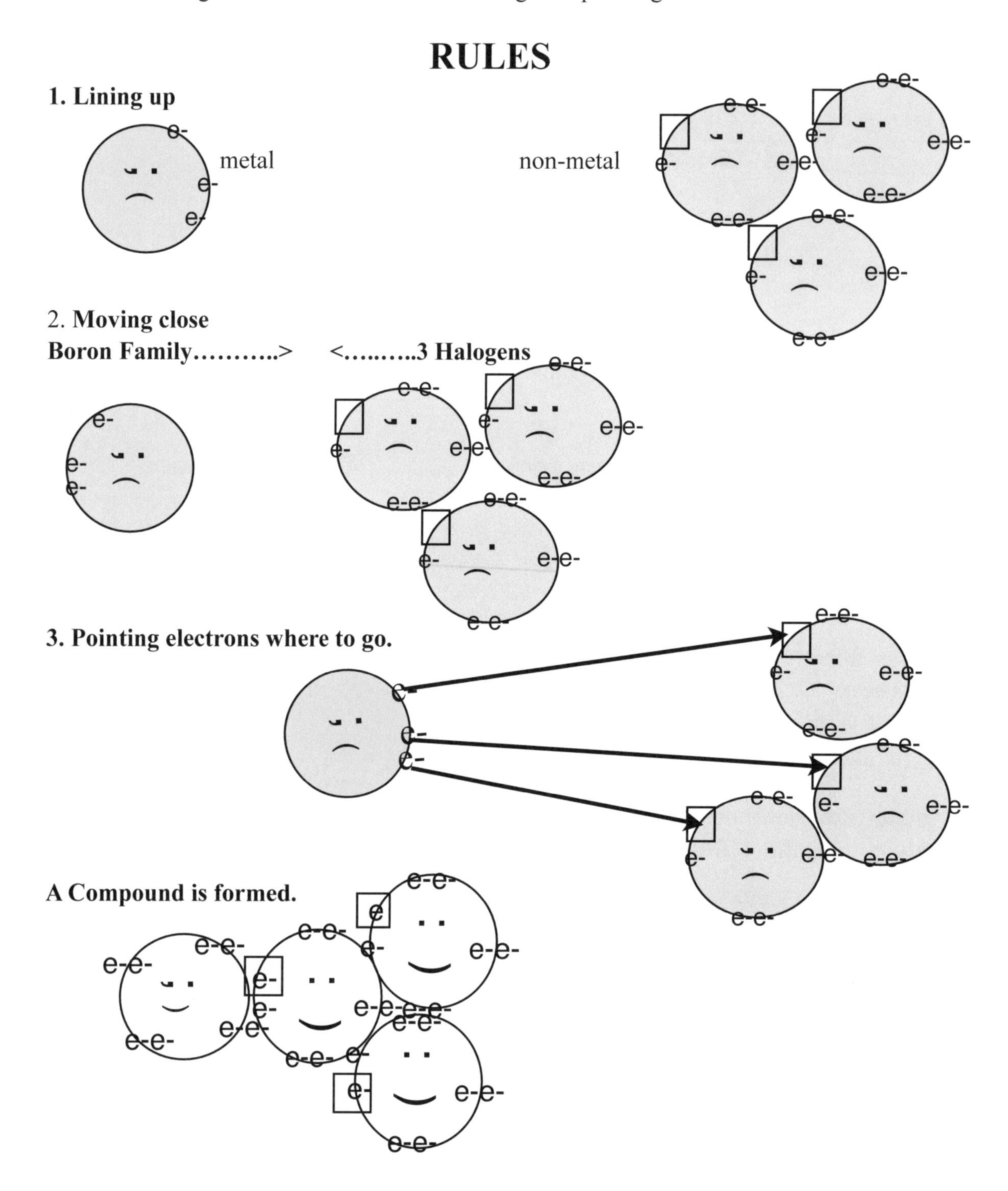

Aluminum said, "It's simple and it will work with each of the Halogens. Just tell the Halogens to come in groups of three. That way each Halogen atom takes one electron from you. The three Halogens take all 3 of your elect

Look at the diagram. The diagram shows that the Boron Family elements each have 3 electrons to get rid of. The arrows show where the electrons go.

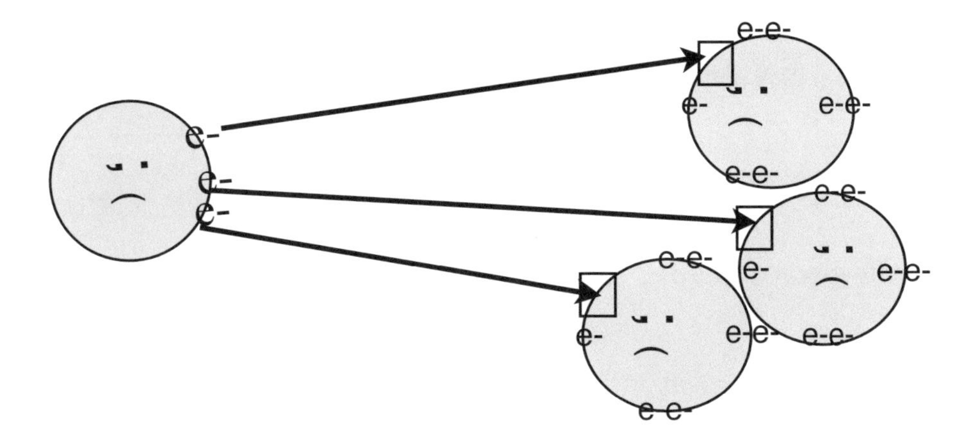

The three Halogens each take one electron. All three Halogen atoms get complete energy levels, and become Happy Atoms. When we get rid of our three electrons, our complete energy level which was hidden becomes our Outside Energy Level. Together we form a new compound and become Happy Atoms."

Boron said, "How this works makes sense, and it's a great idea, and we can make so many different compounds with the Halogens. However, Aluminum I'm still worried that because I am different it won't work this time. I've told you that my protons make me share my electrons instead of giving them away like you.

Aluminum said, "Stop worrying, Boron. It's OK that you are more complicated. You were able to form a compound before and it will work again. Think positive. Our types of bonding are different. That doesn't matter. You bond by sharing electrons. It's covalent bonding. My bonding happens by giving away electrons. It's Ionic Bonding. How your bonding works is not what's important right now. Forget about your covalent bonding. Forget about everything you have different: forget that you're a metalloid and have to make the choice to be a metal; forget that your outside energy level is complete with only 2 electrons. Focus on what does matter, which is that you want to become a Happy Atom. That's your goal. You did it before and you'll do it this time too. I promise you."

They finally arrived at 7th Street, Boron looked around, and decided to make a compound with Fluorine. So they stopped by Fluorine's house, and he knew all about this new way of becoming a Happy Atom.

Boron and Fluorine Become Happy Atoms

So they lined up properly in the same way it was always done.

Boron, acting as a **metal,** with electrons to get rid of was **on the left.**

3 Fluorine atoms, the **non-metals**, who needed to get electrons lined up **on the right**.

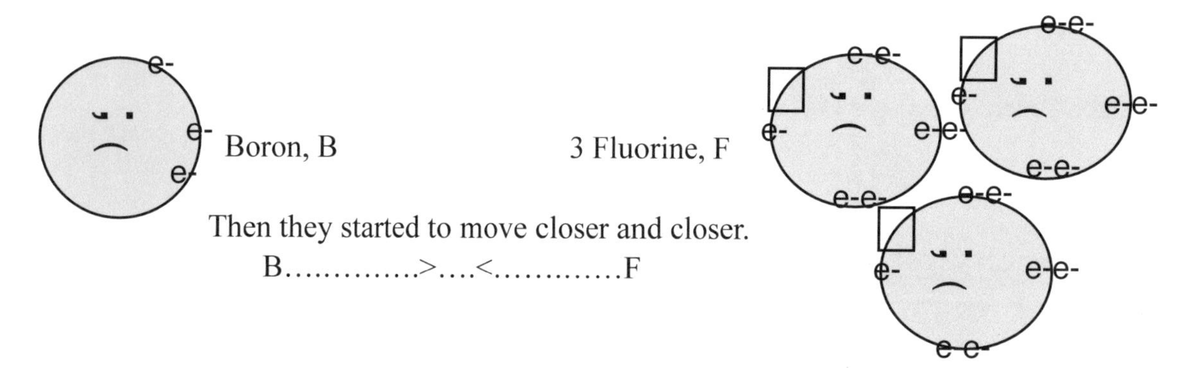

Then they started to move closer and closer.
B.............>....<............F

The arrows show Boron's electrons aimed at the empty space in the three Fluorine atoms.

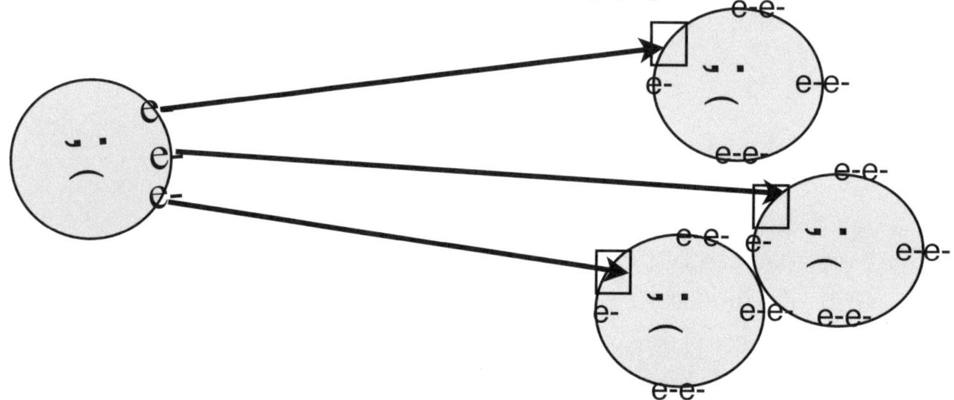

Suddenly when Boron was close enough his three electrons moved into the empty space in each of the three Fluorine atoms' Outside Energy Levels. Fluorine's 3 atoms became happy immediately as they each got the one electron they needed to be a Happy Atom. Boron also became a Happy Atom because his complete energy level was freed up once his 3 electrons became part of Fluorine. Yes, just as Aluminum predicted, I'm complete with only 2 electrons. Then Boron turned to Fluorine and whispered, "Don't worry. We may be sharing electrons but we've become a compound. Better still, we have become Happy Atoms. Look at our smiling happy faces."

Boron and Fluor*ine* have **a new name**...............Boron Fluor***ide***
Their symbols changed to a **formula**.....................BF_3

*Notice, Fluorine changed the ending of his name from **ine** to ide.*
*Fluor**ine** became Fluor**ide.** Also, the subscript 3 under the symbol F_3 means it took 3 Fluorine atoms to make this compound. Count them*

Fluorine said, "One more thing my family wants our compound to be called **Tri**fluor**ide**. You know how fussy my family is. They said, "If Boron needed 3 of us, to make the compound, we deserve to have Tri in our name (Tri means 3). Trifluoride means 3 Fluorine atoms."So this compound is sometimes called Boron Trifluoride."\

Boron combining with Fluorine can be written with words and diagrams:

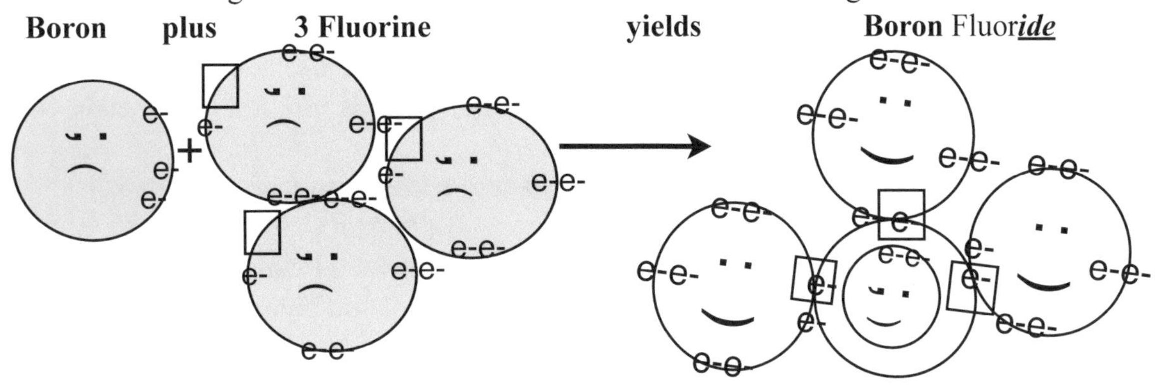

Looking again at the diagram of the Happy Atoms, Boron remembered Aluminum's prediction. Indeed his complete energy level had only 2 electrons in it, and he was happy.

Professor Terry decided to review with Guy why having 2 electrons could make the atom complete. Guy look at your Bohr models for Atomic Numbers 3, 4, and 5. Notice these elements only have 2 energy levels, and we can write them a short way:

Lithium 2) 1, Beryllium 2) 2, Boron 2) 3.

These are the only elements that end up with a complete Outside Energy Level with only 2 electrons in it when they form compounds. Look! Take away the outside electrons and this is what is left:

Lithium 2), Beryllium 2), Boron 2).

Remember the energy level next to the nucleus is complete with 2 electrons. Now Guy, let's continue watching the elements form compounds.

Boron was beyond happy and proud to be adding a compound to his Chemical Family. He was bound and determined to form a compound with all the Halogens while he was at 7th Street. Astatine opted out and left because it was a rare occasion that he wanted to form compounds. Then there were just three more Halogens left: Chlorine, Bromine, and Iodine. As you remember, Boron had already formed a compound with Fluorine. Then, Boron met with all three of these elements at the same time, reviewed the rules and sent them away to come back with three of their atoms.

Chlorine, Bromine, and Iodine, returned. Each brought with them 3 atoms. Boron reminded them of the routine. Then each one lined up in turn to form a compound.

The 3 Halogens stood on the right, Boron on the left side. They moved closer and closer and closer. When they reached that magic distance, each of Boron's electrons moved into the empty spaces in the Halogen's Outside Energy Level. A new compound was formed. Boron repeated this with each of the Halogens and in the end there were 3 more of compounds formed in no time flat. They all became Happy Atoms.

Boron plus 3 Chlorine made Boron Chlor**ide** BCl_3
Boron plus 3 Iodine made Boron Iod**ide** BI_3
Boron plus 3 Bromine made Boron Brom**ide** BBr_3

Since there are 3 Halogens forming these compounds, the Halogens prefer these compounds to be called Boron Tri-Chloride, Boron Tri-Iodide, and Boron Tri-Fluoride. Knowing that Tri means 3, this makes sense.

Aluminum seeing all the compounds Boron had made, decided to do the same thing. "I made compounds with Bromine before. Astatine makes very few compounds. So I'll just send for Iodine, Chlorine, and Fluorine."

These are the compounds formed by Aluminum with the rest of the Halogens

Aluminum plus 3 Chlorine made Aluminum Chlor**ide** $AlCl_3$
Aluminum plus 3 Iodine made Aluminum Iod**ide** $Al\ I_3$
Aluminum plus 3 Fluorine made Aluminum Fluor**ide** AlF_3

The Halogen's fussy Aunt came by and said, to Aluminum, "Show a little respect for us Halogens. I told you these compounds should be called Aluminum Tri-Chloride, Aluminum Tri-Iodide and Aluminum Tri-Fluoride because it wouldn't be a compound if we Halogens didn't come in groups of three." And so that's what you often see these compounds called.

QUIZ

Let's see you make a compound with Aluminum and Iodine.

Guy looked for Aluminum, (Al) and Iodine, (I) on the Periodic Table (Page 85).

Aluminum is in **Group 3A,** and it has **3 electrons to give away**.

Iodine is in **Group 7 A,** and it **needs 1 more electron** to be complete with 8 electrons in his Outside Energy Level.

Guy said, "Iodine can only take 1 electron. Aluminum needs to give away 3 electrons. The only way they can make a compound is for Iodine to come with 3 Iodine atoms, and each one will take one electron from Aluminum. So the formula is $Al\ I_3$, and I remembered that we never write 1 as a subscript.

Here's a word equation and diagram for Aluminum combining with Iodine."

Aluminum. + 3 Iodine Yields. Aluminum Iodide

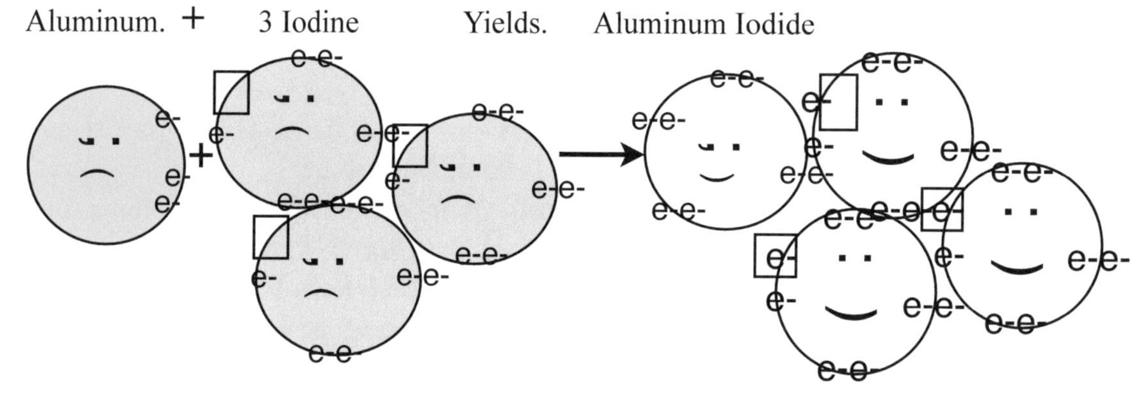

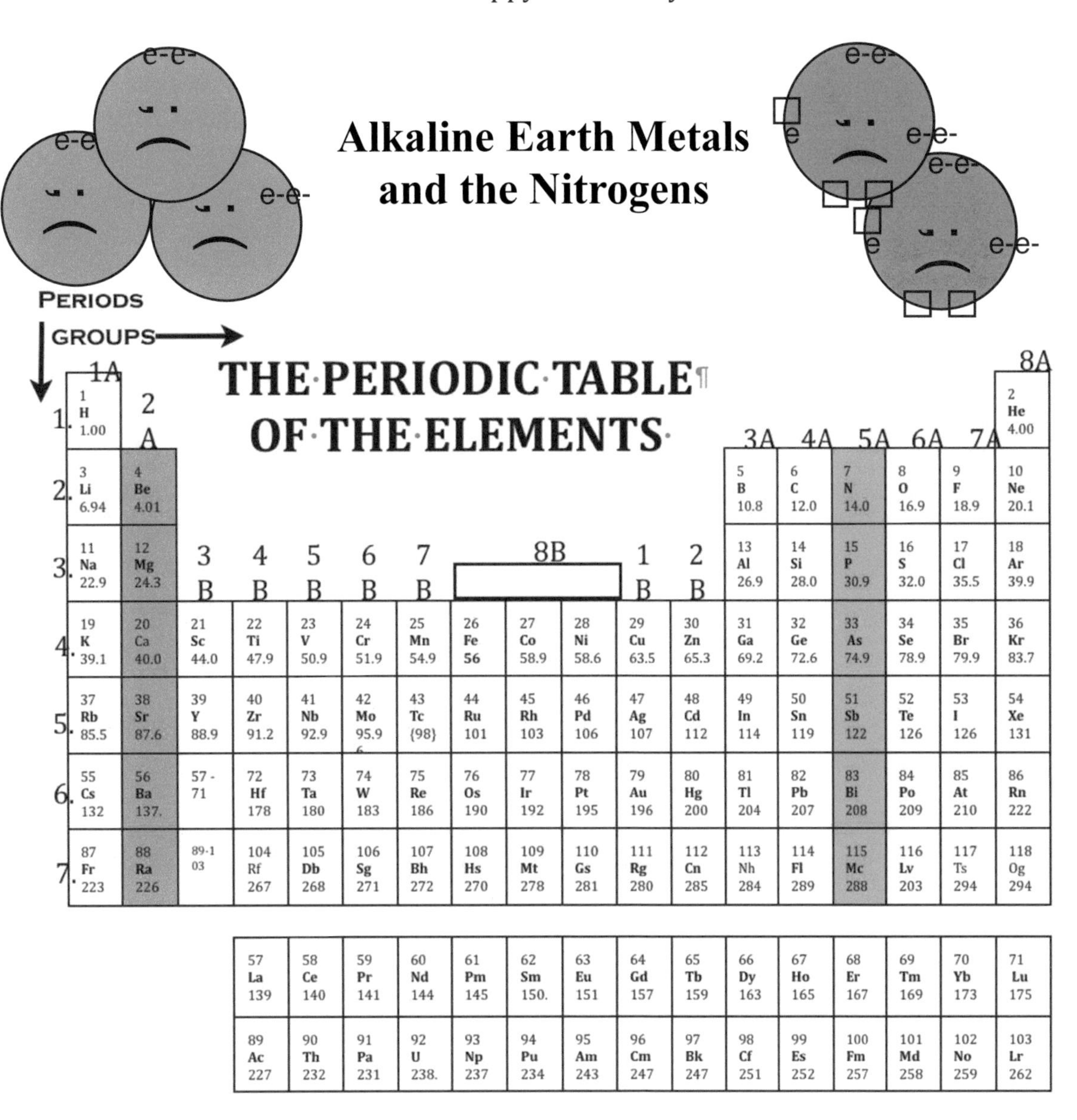

Period	1A	2A	3B	4B	5B	6B	7B	8B			1B	2B	3A	4A	5A	6A	7A	8A
1.	1 H 1.00																	2 He 4.00
2.	3 Li 6.94	4 Be 4.01											5 B 10.8	6 C 12.0	7 N 14.0	8 O 16.9	9 F 18.9	10 Ne 20.1
3.	11 Na 22.9	12 Mg 24.3											13 Al 26.9	14 Si 28.0	15 P 30.9	16 S 32.0	17 Cl 35.5	18 Ar 39.9
4.	19 K 39.1	20 Ca 40.0	21 Sc 44.0	22 Ti 47.9	23 V 50.9	24 Cr 51.9	25 Mn 54.9	26 Fe 56	27 Co 58.9	28 Ni 58.6	29 Cu 63.5	30 Zn 65.3	31 Ga 69.2	32 Ge 72.6	33 As 74.9	34 Se 78.9	35 Br 79.9	36 Kr 83.7
5.	37 Rb 85.5	38 Sr 87.6	39 Y 88.9	40 Zr 91.2	41 Nb 92.9	42 Mo 95.9	43 Tc (98)	44 Ru 101	45 Rh 103	46 Pd 106	47 Ag 107	48 Cd 112	49 In 114	50 Sn 119	51 Sb 122	52 Te 126	53 I 126	54 Xe 131
6.	55 Cs 132	56 Ba 137.	57 - 71	72 Hf 178	73 Ta 180	74 W 183	75 Re 186	76 Os 190	77 Ir 192	78 Pt 195	79 Au 196	80 Hg 200	81 Tl 204	82 Pb 207	83 Bi 208	84 Po 209	85 At 210	86 Rn 222
7.	87 Fr 223	88 Ra 226	89-1 03	104 Rf 267	105 Db 268	106 Sg 271	107 Bh 272	108 Hs 270	109 Mt 278	110 Gs 281	111 Rg 280	112 Cn 285	113 Nh 284	114 Fl 289	115 Mc 288	116 Lv 203	117 Ts 294	118 Og 294

57 La 139	58 Ce 140	59 Pr 141	60 Nd 144	61 Pm 145	62 Sm 150.	63 Eu 151	64 Gd 157	65 Tb 159	66 Dy 163	67 Ho 165	68 Er 167	69 Tm 169	70 Yb 173	71 Lu 175
89 Ac 227	90 Th 232	91 Pa 231	92 U 238.	93 Np 237	94 Pu 234	95 Am 243	96 Cm 247	97 Bk 247	98 Cf 251	99 Es 252	100 Fm 257	101 Md 258	102 No 259	103 Lr 262

Become Happy Atoms
Chapter 9

Then Professor Terry prepared Guy for the next kind of compound formation. “What is coming next is the most challenging way elements become compounds. Guy, you will have to be on your toes and give it all your attention. It’s important for you to learn these challenging types of compound formations. Let’s go back to where we landed and set up a camp site. We need to rest for a little while before the next compound formations begin. Let’s go now so we have a lot of rest time. I predict, Calcium will be the first element to come for help with that difficult compound formation.”

****

Calcium got into the spirit of making Happy Atoms. He and his fellow Alkaline Earth Metals had formed many compounds while on 6th Street and 7th Street adding many Happy Atoms to their Alkaline Earth Metal family. 2nd Street was extremely happy.

The Alkaline Earth Metals hadn't yet visited 5th Street. Calcium wanted to try to make compounds on 5th Street, but it surely looked impossible. Definitely it was not as easy as just asking the Halogens to come in pairs to form compounds. No, it was much more complicated. Calcium decided he needed help, and he was told that Professor Terry and Guy had set up a base camp in the wooded area behind 6th Street. So, Calcium readied his amber colored balloons, slid into the bubble, and off he flew to 6th Street.

******

Back on 6th Street, Guy and Professor Terry had just finished putting the camp site in order. They moved their camp chairs into the shade of an old oak tree to get some rest. Guy got comfortable in his chair, and before long his mind slipped off into dream land. One happy thought after another drifted into his mind. Memories of his nights on the mountain searching the night sky for his favorite constellations and dreaming about the planets whirling around in space were really special. Then Wish Star floated into his dream just as he had appeared that night on the mountain when Guy recited the little poem "Star light, star bright first star I see tonight. I wish I may I wish I might have the wish I wish tonight. Without even wishing Wish Star appeared.

I didn't even have a chance to make a wish, and look what a fantastic summer I've had—meeting Professor Terry with her magic Periodic Table; sliding down the secret tunnel into the fantasy world of Periodic Table Land; flying around Periodic Table Land in a bubble; meeting the elements and the Chemical Families; what fun drawing Bohr Models for the elements.

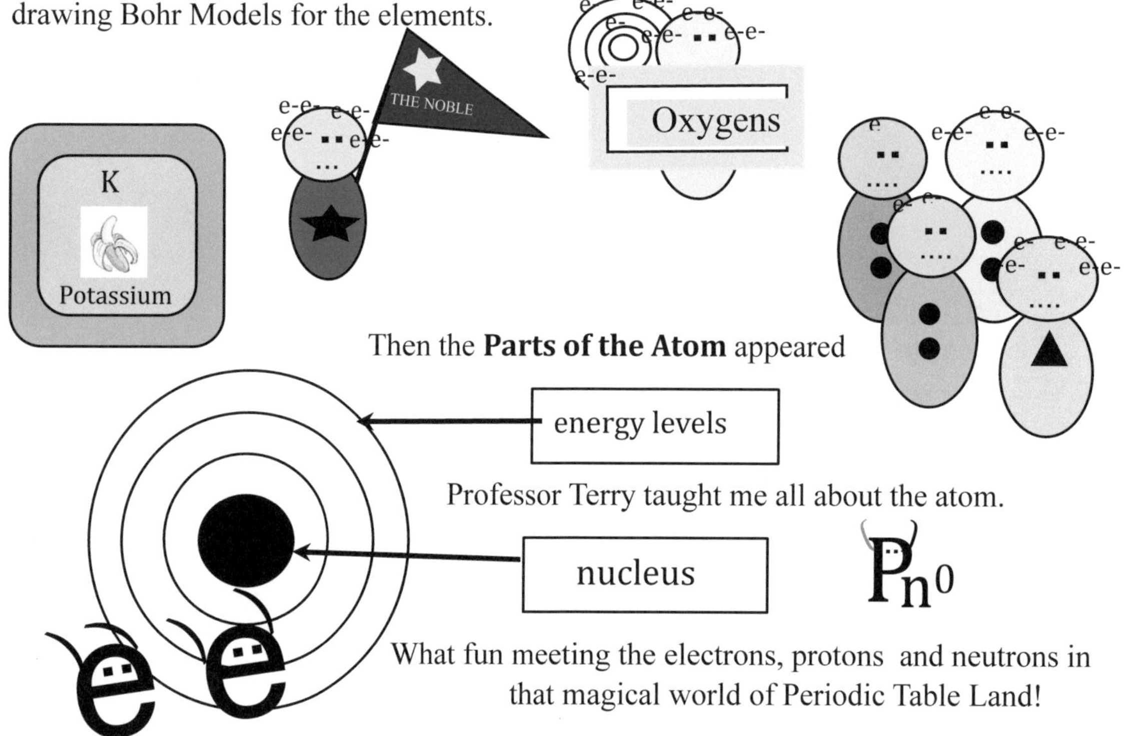

Then the **Parts of the Atom** appeared

Professor Terry taught me all about the atom.

What fun meeting the electrons, protons and neutrons in that magical world of Periodic Table Land!

Guy loved the world around him and he was waking up. He tried unsuccessfully to imagine the whole invisible world that he was learning about existing beneath the surface of the world he loved. It was hard to imagine it truly existing because he couldn't see it. But it did exist. He fell back dreaming about all the magical experiences he had all summer while learning all about chemistry. Then he began to feel a cool breeze brushing across his face. Guy slowly emerged from his dream and returned to the real world. He was sitting in a comfortable chair underneath the oak tree and there was Professor Terry. Guy turned told her about his dream. "I saw so many of the characters that have made my summer so special."

Professor Terry said, "It's good to sit back and remember all that you've learned so far. It was learning all that chemistry which prepared you to understand how compounds were formed and how elements became Happy Atoms."

"I think the most important thing I learned so far about compounds is that the elements are able to create something new and different just by achieving a complete Outside Energy Level. Remember when we were in the bubble that made us invisible and we saw Sodium and Chlorine become a beautiful white crystal, salt. This magic transformation came about just because Sodium gave Chlorine his one electron and Chlorine's Outside energy Level became complete. Sodium's became complete too."

Guy continued, "It was really exciting to see two atoms combine when the non-metal needed the exact same number of electrons that the metal had to give away. **Group 1A** elements had 1 electron to give the elements in Group 7 A who needed 1 electron. **Group 2A** elements had the 2 electrons to give the elements in Group 6A who need 2 electrons. The same was true for **Group 3A** elements. They had 3 electrons to give the elements in Group 5 who needed exactly 3 electrons to create a new compound. It was simple and understandable."

"It was more amazing to see a compound formed when it took more than one atom of an element to make the exchange of electrons possible. The Alkali Metals were able to make compounds with the elements in Groups 5A, 6A and 7A which are Streets 5, 6, and 7. The Alkaline Earth Metals only made compounds on 6th and 7th Streets and the Boron Family only made compounds on 5th Street and 7th Street."

Then Guy asked, "Why didn't the Boron family continue and form compounds on 6th Street? I have the same question about why the Alkaline Earth Metals hesitated to go to 5th Street to form compounds."

Professor Terry was ready with a response. "The streets they had not tried were going to demand a more complicated exchange of electrons. You will have to watch closely to understand the compound formation that will happen when they decide to form compounds on these two streets."

"Guy, I see Calcium approaching. He probably has something bothering him about that same complicated compound formation that we were just talking about. Let's see what he has to say."

Professor Terry called to Calcium, "Come on over here and join us, Calcium."

Calcium walked over to where Guy and the professor were sitting. Finding a chair, he pulled it up beside them. Looking frustrated, he explained the difficulty his Alkaline

Earth Metals encountered as they tried to form compounds with Nitrogen and Phosphorus on 5th Street. "We have only two electrons to give away, and the Nitrogen Family needs three electrons, to have a complete Outside Energy Level. We actually tried a couple of times after you left to form a compound, but we got nowhere. We were always one electron short of creating complete Outside Energy Levels. I came here with one question. Is this ever going to work, or are we just wasting our time trying?"

Professor Terry assured Calcium that it was possible for them to form compounds, and become Happy Atoms with Nitrogen and Phosphorus on 5th Street. She agreed to accompany him to 5th Street and see if she could help him figure it out.

Professor Terry said, "Guy, we need to help Calcium. This is one of the difficult compound formations I was telling you about. Let's get started. You have a lot to learn."

They all climbed into the bubble beneath Calcium's lovely amber colored family balloons and flew over the tree tops to the landing strip behind 5th Street.

When they got there. Phosphorus was out in front of his house polishing his lovely brass sign, #15. That was his address on 5th Street; and, as we know, it was also his Atomic Number. Guy piped up, "It's also the number of protons and electrons he has in his atom."

Professor Terry took the lead, as they approached Phosphorus. "Phosphorus," she began, "I hear you are having trouble creating compounds with the Alkaline Earth Metals. I'm here to tell you it is possible."

Phosphorus said, "We tried. It's impossible!"

Professor Terry said, "No, my friend, it's not impossible. It's difficult. I know how to accomplish it. Calcium would like to try again, and I hope you will agree to try once more with my guidance."

Phosphorus said, "It would be nice to make more compounds, and there are a lot of Alkaline Earth Metals to combine with if we could figure out how to do it. If you will help, Professor Terry, I'll give it one more try."

Professor Terry began, "My advice is to begin, as you did, with all the other elements, and then do just one step at a time. The secret is to keep going, adding one more atom of each element until all the atoms have the exact number of electrons needed to make all their Outside Energy Levels complete. In the end, you will be successful in forming a new compound and becoming a Happy Atom."

"Just remember, the secret is to be patient and don't give up until all the atoms involved have complete Outside Energy Levels. Then, the new compound will be formed. I promise you, it will happen. Patience is what this takes. I'll start you off, and then Guy and I will be right here watching. Just yell, if you need help."

Calcium and Phosphorus Become Happy Atoms

"Let's begin!" said Calcium to Phosphorus.

Step #1 Line up.

The atom with electrons to give away is on the left. That's the **metal, Calcium**.
The atom to get electrons is on the right. That's the **non-metal, Phosphorus.**

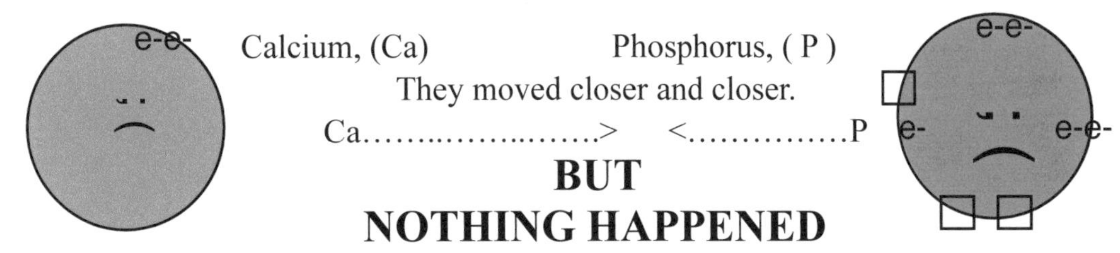

Suddenly they remembered what Professor Terry had said, “Keep going. Add one more atom at a time until the metal gives away all his electrons and the non-metal fills all the empty spaces in his Outside Energy Level. In the end you will make a compound”. That’s what they had to do now. They needed to have the element that did not have the right number of electrons get another atom of that element with more electrons. Phosphorus knew that’s what he had to do. He had to tell Calcium to go get another Calcium atom to give him that extra electron he needed That’s what they did..

Readers, follow the numbers

1. Calcium said, “I have 2 electrons. If I give you these 2 electrons, I will be a “Happy Atom”

2. Phosphorus said, “If I take your 2 electrons, I will not be happy because I will still be missing one electron to have a complete Outside Energy Level. I need one more electron. So go get another Calcium atom to give me one more electron.”

3. Calcium got Calcium atom #2 to come with the needed electron. “Now let’s see happens.”

Ca

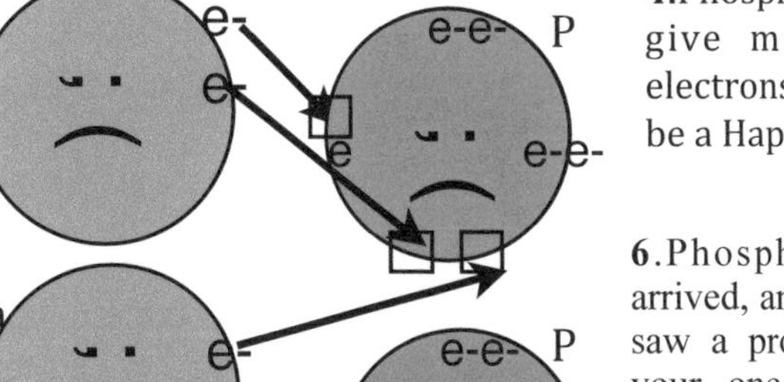

4. Phosphorus said, “Good, give me one of your electrons now, and I will be a Happy Atom.”

5. Calcium #2 said, “No. Then I will be stuck with an extra electron, and I will not be happy. So go get another Phosphorus atom.”

6. Phosphorus atom #2 arrived, and he immediately saw a problem. “If I take your one electron, I will still be missing 2 electrons. I will not be a Happy Atom.”

7. Calcium said, “Problem solved. If I get one more Calcium atom he will have the 2 electrons you need.” So he sent for another Calcium atom and guess what happened?

Ca

8. Phosphorus said, “When this Calcium gives me his two electrons. All the spaces in my Outside Energy Levels will be filled and Calcium will get rid of all his electrons, too. We’ll both be Happy Atoms.” The electrons did their jump, and a new compound was formed. Look at their happy faces.

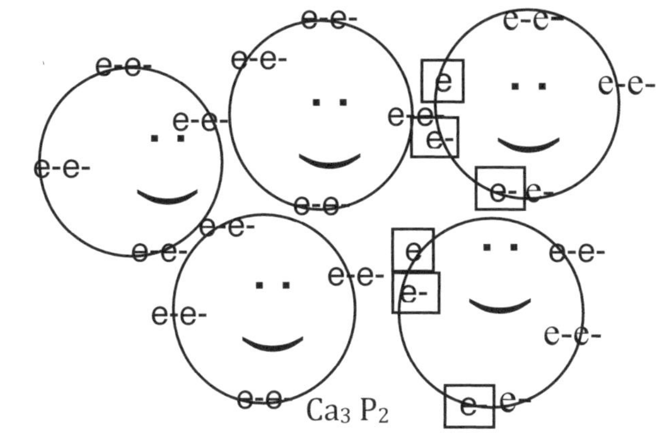

9. Count the number of Calcium and Phosphorus atoms it took to make this compound. There were 3 Ca atoms and 2 P atoms. So the formula is.Ca_3P_2 They are so happy now. Calcium and Phosphorus have become Happy Atoms. A new compound was formed. The process was difficult, but with patience they succeeded.

They have a **new** name.....................Calcium Phosph**ide**

Their symbols changed into a **formula**.........Ca_3P_2

Notice the element that took in the electrons, the non-metal

*changed his name's ending. Phosph**orus** became Phosph**ide***

Now look at the little number under each symbols.

Ca_3P_2. both elements to have complete Outside Energy Levels. That's what the formula Ca_3P_2. for that compounds means.

This is how we write what happened in words and diagrams:

3 Calcium atoms plus 2 Phosphorus atoms yields Calcium Phosphide.

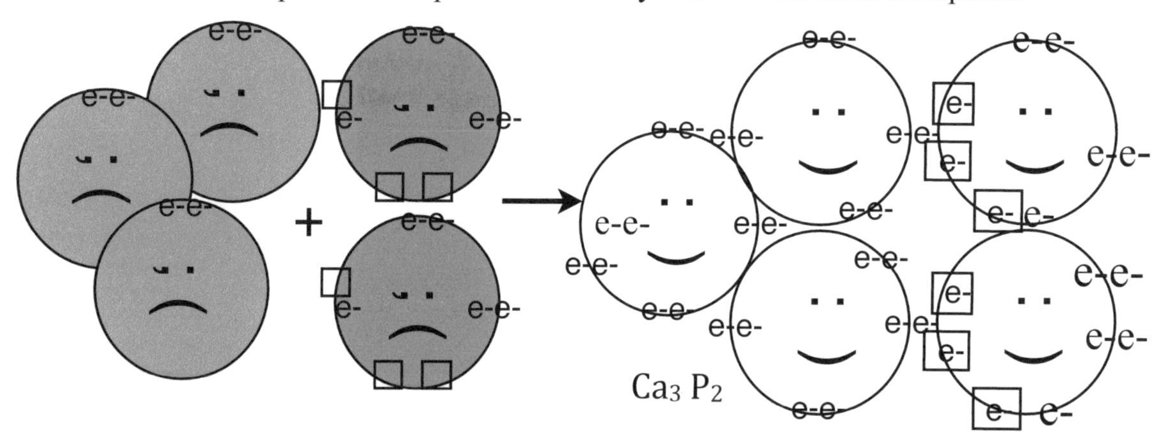

Professor Terry told Guy, "This is called a chemical equation, and we will later learn how to write this equation using symbols and formulas. After that, we will learn how to balance the equation because in nature the reaction is always balanced. This is just a preview of what you will be learning next."

"Tell me again where we get those little numbers we wrote below the elements' symbols in the formula for Calcium Phosphide, $Ca_3 P_2$.?"

Patiently, Professor Terry told him, "We counted how many Calcium atoms it took to make Phosphorus a Happy Atom. Count them Guy."

He said, "It took 3 Calcium atoms."

Professor Terry continued, "That's why we put 3 below the Calcium, Ca_3. Then we counted how many Phosphorus atoms it took to make Calcium a Happy Atom. Count them."

Guy said, "It was 2. I get it now. That's why we put the 2 below the symbol for Phosphorus and wrote P_2. Then we had the whole formula for Calcium Phosphide, Ca_3P_2." He was happy that he understood the meaning of the numbers in the formula.

Professor Terry said, "The most important thing to remember is that Calcium and Phosphorus became a compound when both elements achieved complete Outside Energy Levels."

Then Calcium and Phosphorus came over to thank Professor Terry for her help. "Now, it is possible for all of the Alkaline Earth Metals to form compounds with the elements on 5th Street. We are so happy. Thank you."

Calcium went back to 2nd Street, and brought his whole family back to 5th Street. They were excited to think of making lots of compounds with the Nitrogen family. Their amber colored balloons landed the Alkaline Earth Metals safely on the tarmac behind 5th Street. All the 2nd Street elements: Beryllium, Magnesium, Calcium, Strontium, Barium and Radium piled out of the family bus. Calcium led the way through the forest.

When they got to 5th Street, they saw Nitrogen family's beautiful red flag flapping in the breeze. Calcium saw Phosphorus and negotiated a plan to have his elements form compounds with Nitrogen as well as with Phosphorus.

Calcium went over and spoke to Nitrogen. "With Professor Terry's help, we did it. Phosphorus and I learned how our families could form compounds and become Happy Atoms. My whole family is here to create compounds. Magnesium said he would like to go first and join up with you, Nitrogen. It's time for you to get started."

Magnesium and Nitrogen came forward. The rest of the Alkaline Earth Metals sat nearby to watch the process unfold. That way they would be ready to form their own compounds.

Magnesium and Nitrogen Become A Compound

Magnesium and Nitrogen said, " Let's begin."

2nd Street, **Magnesium**, the **metal** with electrons to give away stood on the left.

5th Street, **Nitrogen**, the **non-metal** missing electrons stood on the right.

"This is the easy part." Calcium and Phosphorus exclaimed.

'I wish the whole process was this easy."

Professor Terry listening decided to give Guy some words of wisdom. "Guy life has it challenges as well. When we come up against difficulties the normal thing to do is to get upset and wish things were not so difficult. I want to tell you Guy it's a waste of time to wish a problem didn't exist. Once you are hit with a problem start immediately to figure out what you can do about it. If a problem can't be fixed learn how you can live with it without being upset. Don't waste your time being unhappy. Calcium acted wisely when he came to me for advice. It's good to look for people who can help. Now it's time to watch Magnesium and Nitrogen in action."

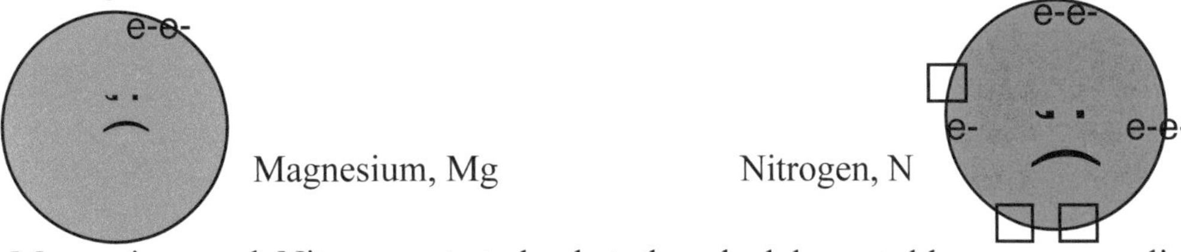

Then Magnesium and Nitrogen started what they had been told was a complicated process. They were ready. They began to move closer and closer

Mg.......................................> <......................................N

AGAIN NOTHING HAPPENED

Nitrogen said, "It's hard to believe that this is possible with an element that has only two electrons to give away. I watched Calcium and Phosphorus do this same exchange with my own eyes. So I know it can be done." Nitrogen was giving himself a pep talk. So he turned to Magnesium and said, "Let's get on with this."

Professor Terry called over from the tree line and said, "I'll be here to help the whole time. Don't worry so much, Nitrogen. You and Magnesium are doing just fine. Remember, don't quit until all the elements get complete Outside Energy Levels, and a new compound is formed."

Nitrogen summoned up all the courage, he could find, and continued forming the compound with Magnesium. Here's what happened one step at a time.

Readers, follow the numbers

1. **Magnesium** said, "I have 2 electrons. If I give away these 2 electrons, I will be a Happy Atom."

2. **Nitrogen** said, "If I take those 2 electrons, I will not be happy because I will still be missing one electron to have a complete Outside Energy Level." Then Nitrogen said "I need one more electron so go get another Magnesium atom."

3. **Magnesium** got another Magnesium atom. **Nitrogen** said, "Give me one of your electrons, and I will be a Happy Atom. **Magnesium** said, "Then I will not be happy. So go get another Nitrogen atom."

4. The **new Nitrogen** atom arrived and he immediately saw a problem. "If I take your one electron, I will still be missing 2 electrons, and I will not be a Happy Atom." **Magnesium** said, "I think if I get one more Magnesium atom, our whole problem will be solved." So he sent for another Magnesium atom. Guess what happened?

5. Magnesium was able to give away his two electrons. All the spaces in Nitrogen's Outside Energy Levels were filled. All the Magnesium atoms were rid of their 2 electrons. The complete hidden energy levels popped up and became a complete a Outside Energy Level for each Magnesium.

6.They are so happy! Magnesium and Nitrogen have now become a compound. Look at their smiling Happy Atom faces. But there is more.

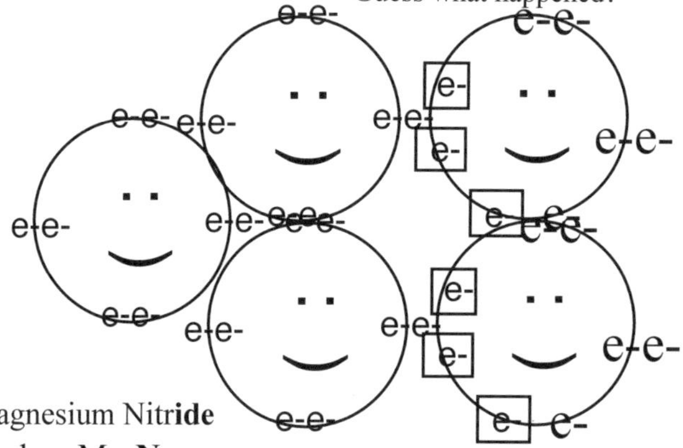

They have a new name ....Magnesium Nitr**ide**

Their symbols changed into a formula...Mg_3N_2

Notice: *The element that took in the electrons changed the ending of his name to ide. Nitr**ogen** became Nitr**ide.** The formula, **Mg_3N_2**. says it took 3 atoms of Magnesium Mg to combine with 2 atoms of Nitrogen N to form the compound.Mg_3N_2*

Guy checked out the formula, and he saw that indeed that's how many atoms it took to make the compound, Magnesium Nitride, Mg_3N_2. There were 3 Magnesium atoms. That's what Mg_3 in the formula was saying. Guy counted 2 Nitrogen atoms.That's what N_2 in the formula was saying. I guess this is the correct formula for Magnesium combing with Nitrogen.

Calcium asked Professor Terry to explain the process to the rest of the Alkaline Earth Metals who will be making compounds with Nitrogen or Phosphorus.

Professor Terry said, "To save time and writing, as I explain this all to you again, I'm going to use AEM, the first letters of your family name (**A**lkaline **E**arth **M**etals). When you hear me say AEM, I want each of you elements to think of your own name when I say or write AEM You are the brown elements, and the metals with electrons to give away. I'll also use NF for the **N**itrogen **F**amily instead of writing out Nitrogen and Phosphorus. When you hear me say or write NF think of the Nitrogen Family element that you will be making the compound with. They are the red elements.

Remember follow the numbers as you read what happens.

The Alkaline Earth Metals (AEM) and The Nitrogen Family (NF) Form Compound

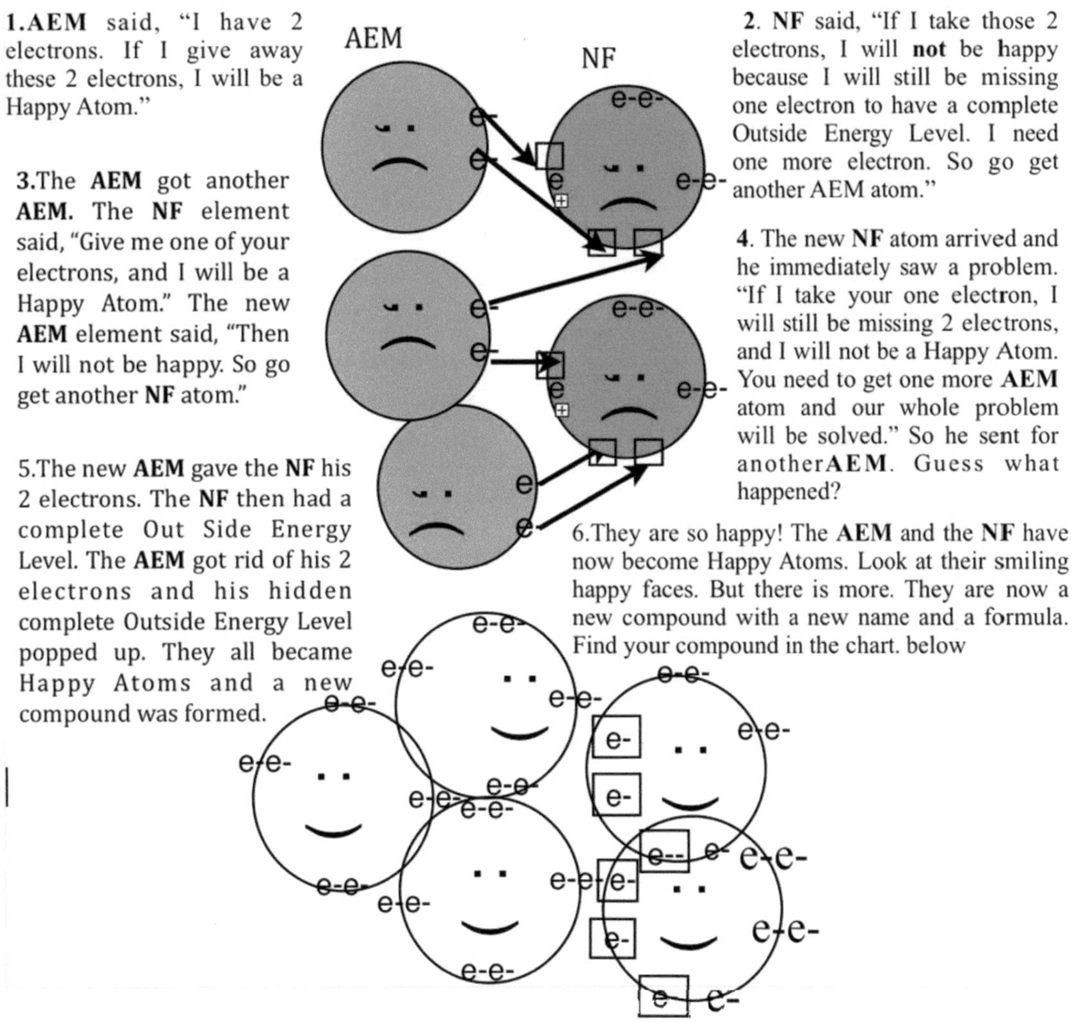

1.AEM said, "I have 2 electrons. If I give away these 2 electrons, I will be a Happy Atom."

2. NF said, "If I take those 2 electrons, I will **not** be happy because I will still be missing one electron to have a complete Outside Energy Level. I need one more electron. So go get another AEM atom."

3.The **AEM** got another **AEM.** The **NF** element said, "Give me one of your electrons, and I will be a Happy Atom." The new **AEM** element said, "Then I will not be happy. So go get another **NF** atom."

4. The new **NF** atom arrived and he immediately saw a problem. "If I take your one electron, I will still be missing 2 electrons, and I will not be a Happy Atom. You need to get one more **AEM** atom and our whole problem will be solved." So he sent for another**AEM**. Guess what happened?

5.The new **AEM** gave the **NF** his 2 electrons. The **NF** then had a complete Out Side Energy Level. The **AEM** got rid of his 2 electrons and his hidden complete Outside Energy Level popped up. They all became Happy Atoms and a new compound was formed.

6.They are so happy! The **AEM** and the **NF** have now become Happy Atoms. Look at their smiling happy faces. But there is more. They are now a new compound with a new name and a formula. Find your compound in the chart. below

:d

when this compound was formed.

3Alkaline Earth Metals + 2 Nitrogen Family elements yields a Compound

3 AEM plus 2 NF Yields A Compound

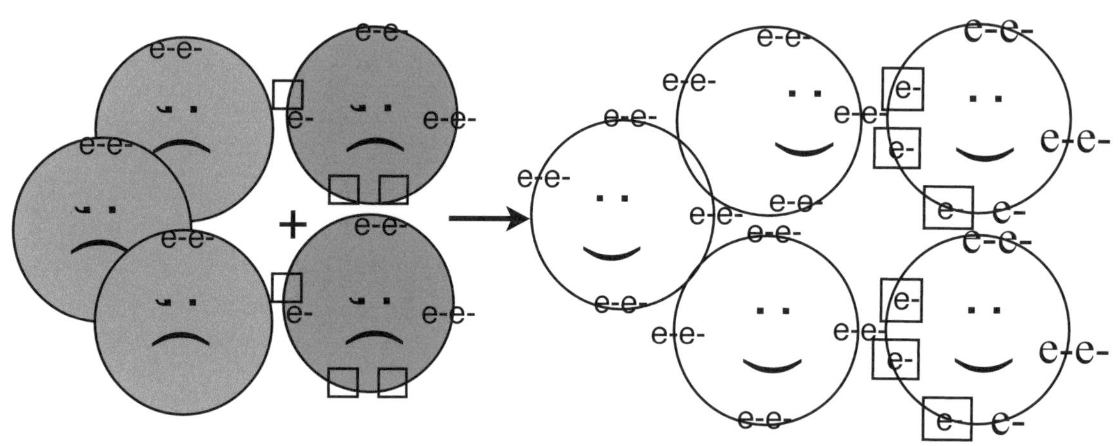

The compound formed depends on which Alkaline Earth Metal (AEM) combined with which Nitrogen Family (NF) element. Below you can see the compounds that were formed just like the example above. It shows what happened when each of the AEM elements: Beryllium, Magnesium, Calcium, Strontium, Barium and Radium joined with a NF element: Nitrogen or Phosphorous. In each compound it took 3 Alkaline Earth Metals to join with 2 Nitrogen Family elements. Thats what the subscripts tell you.

NITROGEN'S COMPOUNDS		PHOSPHORUS' COMPOUNDS	
Beryllium Nitride	Be_3N_2	Beryllium Phosphide	Be_3P_2
Magnesium Nitride	Mg_3N_2	Magnesium Phosphide	Mg_3P_2
Calcium Nitride	Ca_3N_2	Calcium Phosphide	Ca_3P_2
Strontium Nitride	Sr_3N_2	Strontium Phosphide	Sr_3P_2
Barium Nitride	Ba_3N_2	Barium Phosphide	Ba_3P_2
Radium Nitride	Ra_3N_2	Radium Phosphide	Ra

Calcium ran over to thank Professor Terry and Guy for the good advice. "I wanted to give up after my second Calcium atom was not able to make Phosphorus' Outside Energy Level complete, but your words of advice rang in my head, 'Don't give up until the new compound is formed.'. Thank you. I'll try to remember your words of advice even in my personal life when I have problems: *Do not give up.* Just when you think your problem has no solution, along comes a solution and your problem is solved

Professor Terry said, "Sometimes the solution is to learn to live with the problem."

At that moment, Calcium saw Aluminum standing behind the bushes. Aluminum said, "Congratulations Calcium. I was returning home from visiting the Halogens when I saw you doing that complicated electron exchange. I hung around to watch. Now I'm going home to tell Boron I found out how we might possibly make compounds on 6th Street the way you did on 5th Street. We are so much like you. Bye now."

The Boron and Oxygen Family Become HappyAtoms

CHAPTER 10

THE PERIODIC TABLE OF THE ELEMENTS

PERIODS ↓ GROUPS →

Period	1A	2A	3B	4B	5B	6B	7B	8B	8B	8B	1B	2B	3A	4A	5A	6A	7A	8A
1.	1 H 1.00																	2 He 4.00
2.	3 Li 6.94	4 Be 4.01											5 B 10.8	6 C 12.0	7 N 14.0	8 O 16.9	9 F 18.9	10 Ne 20.1
3.	11 Na 22.9	12 Mg 24.3											13 Al 26.9	14 Si 28.0	15 P 30.9	16 S 32.0	17 Cl 35.5	18 Ar 39.9
4.	19 K 39.1	20 Ca 40.0	21 Sc 44.0	22 Ti 47.9	23 V 50.9	24 Cr 51.9	25 Mn 54.9	26 Fe 56	27 Co 58.9	28 Ni 58.6	29 Cu 63.5	30 Zn 65.3	31 Ga 69.2	32 Ge 72.6	33 As 74.9	34 Se 78.9	35 Br 79.9	36 Kr 83.7
5.	37 Rb 85.5	38 Sr 87.6	39 Y 88.9	40 Zr 91.2	41 Nb 92.9	42 Mo 95.9	43 Tc {98}	44 Ru 101	45 Rh 103	46 Pd 106	47 Ag 107	48 Cd 112	49 In 114	50 Sn 119	51 Sb 122	52 Te 126	53 I 126	54 Xe 131
6.	55 Cs 132	56 Ba 137.	57 - 71	72 Hf 178	73 Ta 180	74 W 183	75 Re 186	76 Os 190	77 Ir 192	78 Pt 195	79 Au 196	80 Hg 200	81 Tl 204	82 Pb 207	83 Bi 208	84 Po 209	85 At 210	86 Rn 222
7.	87 Fr 223	88 Ra 226	89-103	104 Rf 267	105 Db 268	106 Sg 271	107 Bh 272	108 Hs 270	109 Mt 278	110 Gs 281	111 Rg 280	112 Cn 285	113 Nh 284	114 Fl 289	115 Mc 288	116 Lv 203	117 Ts 294	118 Og 294

57 La 139	58 Ce 140	59 Pr 141	60 Nd 144	61 Pm 145	62 Sm 150.	63 Eu 151	64 Gd 157	65 Tb 159	66 Dy 163	67 Ho 165	68 Er 167	69 Tm 169	70 Yb 173	71 Lu 175
89 Ac 227	90 Th 232	91 Pa 231	92 U 238.	93 Np 237	94 Pu 234	95 Am 243	96 Cm 247	97 Bk 247	98 Cf 251	99 Es 252	100 Fm 257	101 Md 258	102 No 259	103 Lr 262

Aluminum went over to visit Boron in his sparkling clean house. He told Boron, "I just happened to be passing by 5th Street on my way home from visiting the Halogens. What I saw was amazing. It was awesome to watch how Calcium went about forming a compound with Phosphorus. The exchange of electrons seemed like it would go on forever. Finally Phosphorus and Calcium both achieved complete Outside Energy Levels creating a new compound. You should have seen them become Happy Atoms."

Aluminum thought this could never be done saying, "If I hadn't seen it with my own eyes, I would not have believed it was possible. How could an element with only 2 electrons satisfy an element that needed 3 electrons. But it happened. You know if we were to try to make compounds with the elements on 6th Street, it would involve almost the same exchange of electrons as I observed Calcium and Phosphorus doing, just in reverse. We have 3 electrons to give away, and Oxygen can only take 2 electrons.

Boron gave it some thought and then said, "Those elements on 2nd Street really do have only 2 electrons to give away. The Nitrogen Family elements need 3 electrons. It doesn't seem possible that they could get complete Outside Energy Levels. But you saw them become Happy Atoms. Our situation is almost the same. Thinking out loud Boron said, “We have 3 electrons to give away, and the Oxygen family can only take 2 electrons. It does sound like you’re right. If they were able to become Happy Atoms, we can too.” After contemplating the whole idea Boron finally said, “I say, if you’re sure of what you saw, it’s worth trying. Let’s do it.”

Aluminum told Boron, “Here’s my plan. I’ll go over to 6th Street first and give it a try. If it works, I’ll send for you.” Aluminum slid into the bubble beneath their family’s shiny, silver balloons, pointed his magic wand upward, and over to 6th Street he flew. When he got out of the bubble, he was surprised to see Professor Terry and Guy there too.

Professor Terry said,”Guy and I were discussing with Oxygen that awesome event that Nitrogen had pulled off with Calcium.”

Aluminum said, “I was there too, and it was spectacular.” It inspired me to think about making compounds here on 6th Street. Oxygen, I have 3 electrons to give away, and I know you can only take 2 electrons. However after watching how Calcium worked it out, I’m sure we can create a compound together. We need to try.”

Oxygen said, “Professor Terry already told me about the complicated compound formation that Calcium and Phosphorus had managed to make. She was telling me that she was sure it would work for my family and yours. I am willing to try.”

Professor Terry and Guy asked if they could watch from their bubble which would keep them invisible and out of the way. “Just remember what I told you, Oxygen. The secret of success is ‘Never give up until the new compound is formed even if it seems to go on forever.’ The idea is to keep adding other atoms of your elements, until Aluminum gets rid of all his electrons, and the Oxygen atoms fill up all their empty spaces. It’s then that all your atoms will have complete Outside Energy Levels and you will become Happy Atoms. You will be a new creation, a compound.”

Aluminum was glad that Professor Terry would be nearby. Aluminum looked around trying to find Oxygen. Oxygen had already scooted to the other side of his house. When Aluminum found Oxygen, Oxygen said, “There’s no reason we can’t enjoy the beautiful sunshine while we attempt this complicated compound formation.”

Professor Terry and Guy had already set up their observation post in Oxygen’s yard. Then the action began.

Aluminum, the metal, with electrons to give away stood on the left.
Oxygen, the non-metal looking to take in electrons stood on the right.

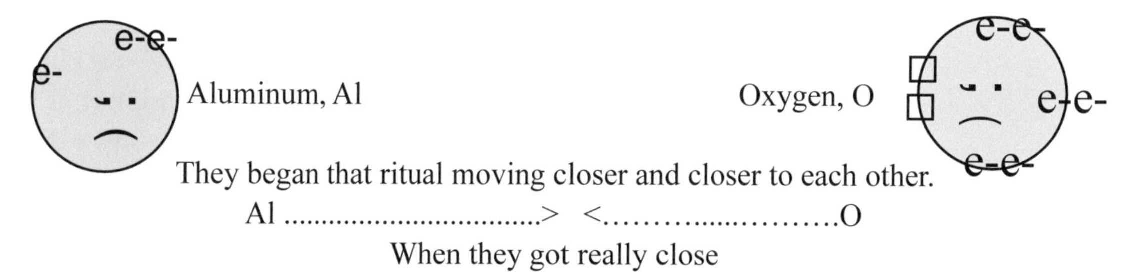

They began that ritual moving closer and closer to each other.

Al ....> <……….....………O

When they got really close

NOTHING HAPPENED

Here's how the compound formation happened step by step in slow motion.
Readers, follow the numbers

1.**Oxygen** said, "From my point of view, I don't see any problem for us becoming Happy Atoms. You have enough electrons to complete my Outside Energy Level."

2. "The problem is this," said **Aluminum**. "I can't give you those 2 electrons unless I'm able to also give you my third electron. Unless I get rid of all my electrons, my complete Outside Energy Level will not rise to the surface. You need to get another Oxygen atom. He can take my extra electron."

3. So **Oxygen** went off and found a second Oxygen atom.

4. **Aluminum** was all excited when the second Oxygen atom arrived, and said. "OK, now we can get started. The first Oxygen gets two of my electrons, and Oxygen #2 gets my third electron. Now I'll be a HappyAtom."

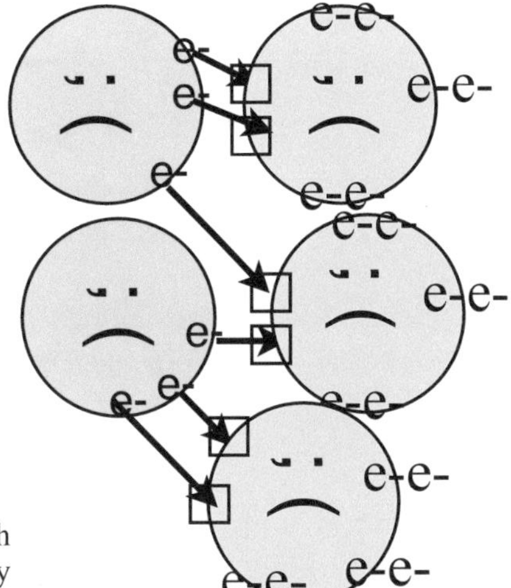

5. The second **Oxygen** atom said, "Not so fast my friend, Aluminum! If I take your one electron, then you will be a Happy Atom; but I will still be missing one electron to have a complete Outside Energy Level. Now you have to get another Aluminum atom."

6. So **Aluminum** came back with the 2nd Aluminum atom. Suddenly when he looked over the situation, he was rather exasperated. "Now we're back where we started. If I give you just one electron, I still have 2 electrons left to give away."

7. **Oxygen** said, "No, it's not the same this time. I see an end to this. I just need to get one more atom of Oxygen and he can take those two extra electrons you have. Then we will all have complete Outside Energy Levels. We will all be Happy Atoms." To everyone's delight it happened. They all became Happy Atoms.

At that precise moment, when the last two electrons jumped from Aluminum to Oxygen's Outside Energy Level a new compound was formed. Off went that joyful sound! A new compound was formed, and here's what the Happy Atoms looked like.

They have a new name ....Aluminum Ox**ide**
Their symbols changed into a formula ....$Al_2\mathbf{O_3}$

In the formula $Al_2\mathbf{O_3}$ the subscript 2 in $\mathbf{Al_2}$ says that 2 atoms of Aluminum combined with Oxygen. The subscript 3 in $\mathbf{O_3}$ Says there are 3 atoms of Oxygen combining with Aluminum to make the compound Aluminum Oxide, $Al_2\mathbf{O_3}$

This is how we write what happened in words and diagrams:

2Aluminum plus 3 Oxygen **yields** Aluminum Oxide

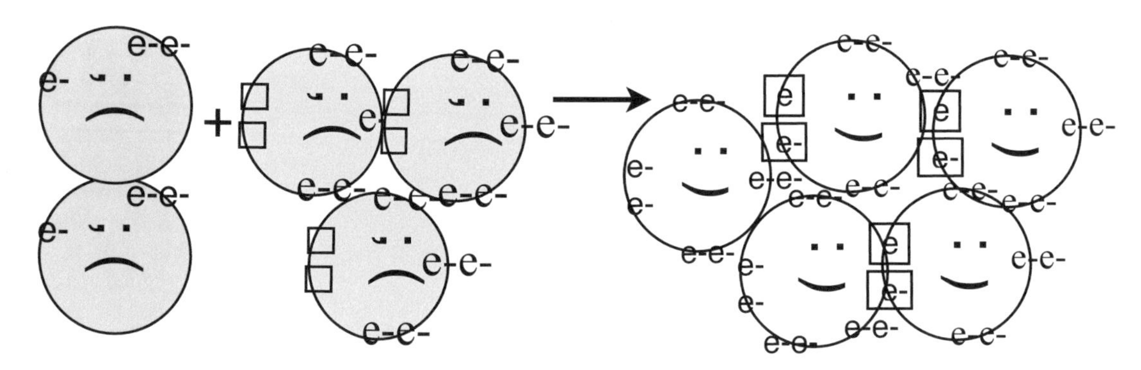

Now that Aluminum proved this could be done, Boron knew he could make compounds with Oxygen too. As a matter of fact Aluminum and Boron can form compounds with both Oxygen and Sulfur and they did. Boron said, "Thank you Aluminum."

Boron Forms a Compound With Sulfur

Since Boron was over on 6th Street, he decided to drop into house #16, the home of Sulfur. Sulfur said,"You're one of the elements from 3rd Street. I can tell by the 3 electrons you have on your head and your gray color." Sulfur invited Boron in to talk. After exchanging pleasantries, he asked, "Did you come for more than a social visit?"

Boron explained, "Yes, I'm here because. Aluminum, a member of my family, and Oxygen, the head of your Family found a way to create compounds together, and I'm interested in seeing if you and I could make a compound too. Are you up for it?"

Sulfur said, "I thought you would never ask. Sulfur observed both their atoms for a while, and then Sulfur said, "Actually, it looks impossible to me. I don't see how both of us could end up with complete Outside Energy Levels."

Boron assured Sulfur that although this kind of compound formation is difficult, it can be done. "The best thing to do is take it one step at a time and never give up until each atom has all the electrons it needs. 'Never giving up' is the secret of success."

Boron said, "Believe me, Aluminum and Oxygen succeeded in making a compound together. So let's give it a try."

Boron and Sulfur stood across from each other, ready to begin.They knew the metal Boron should be on the left and the non-metal, Sulfur on the right.

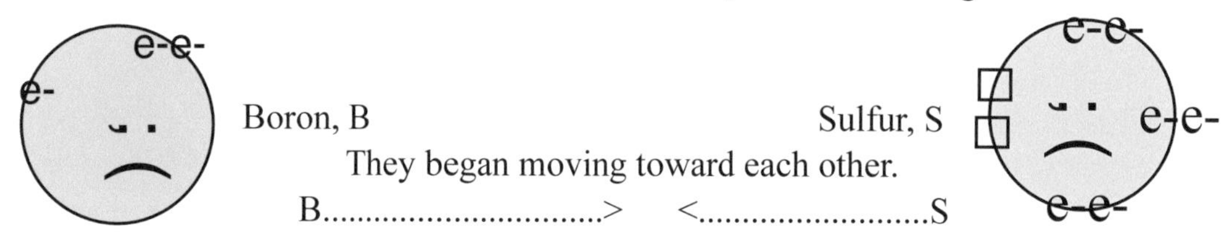

They began moving toward each other.

B...............................> <..........................S

When they got really close..........

NOTHING HAPPENED

Readers follow the numbers

Here's how the compound formation happened step by step in slow motion.

1.**Sulfur** said, " Now that I see you up close, from my point of view, I don't see any problem for us becoming Happy Atoms. You have more than enough electrons to complete my Outside Energy Level."

2. "The problem is this," said **Boron** "I can't give you those 2 electrons unless I'm able to get rid of my third electron. Unless I get rid of all my electrons, my complete Outside Energy Level will not rise to the surface, and I won't be happy. You need to get another Sulfur atom to take my extra electron."

3. So **Sulfur** went off and found a second Sulfur atom.

4. **Boron** was all excited when the second Sulfur atom arrived, and said."OK, now we can get started. The first Sulfur atom gets 2 of my electrons and the second gets my other electron. Now I'll be a Happy Atom."

5. The second **Sulfur** atom said, "Not so fast my friend, Boron! If I take your one electron then you will be a Happy Atom, but I will still need one more electron to have a complete Outside Energy Level. Now you have to get another Boron atom."

6. **Boron** came back with the second Boron atom. Suddenly when he looked over the situation he was rather exasperated. "Now we're back where we started. If I give you just one electron. I will be stuck with 2 electrons."

7, **Sulfur** said, "No, it's not the same this time. I can end to this. I just need to get one more atom of Sulfur and he can take those two extra electrons you have. Then we will both have complete Outside Energy Levels. We will both be Happy Atoms."

8. Sulfur found another Sulfur atom who came and took Boron's other two electrons, and they all became Happy Atoms. Look at their smiling faces..

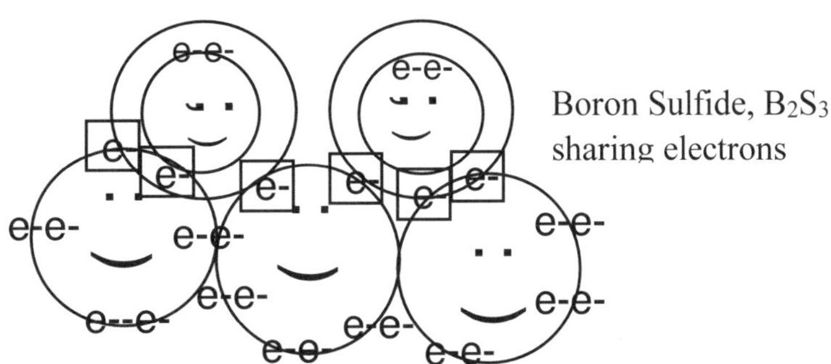

Boron Sulfide, B_2S_3 sharing electrons

They are so happy now. They have become a compound. Forming this compound was much more involved than an ordinary exchange of electrons. The structure of Boron's atom caused him to be covalent. This meant he had to share electrons instead of giving them away. The good outcome was that in the end a compound was formed which was what they hoped would happen. The compound's name and formula were the same as in an ordinary exchange of electrons. It didn't matter that Boron was sharing, not giving Sulfur the electrons he needed. All that mattered was that a new compound was formed and the new name and formula were the same as Boron and Sulfur had hoped they would be. This made Boron and Sulfur so very happy.

This is how we write what happened in words and diagrams:

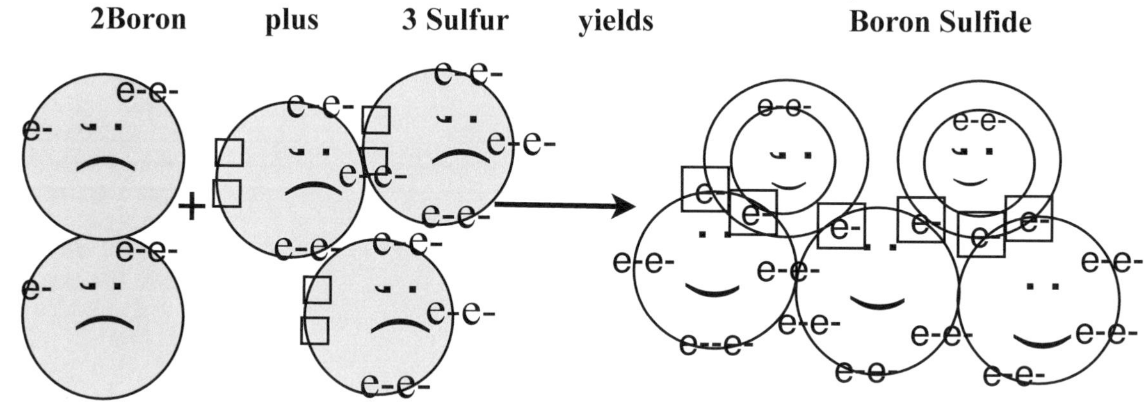

****

After that Aluminum decided to make Happy Atoms with Sulfur, this is what happened: Aluminum Sulfide was formed with the formula, Al_2S_3

Then Boron decide to combine with Oxygen and here's what happened Boron Oxide was formed with the formula, B_2O_3..

Forming both these compounds took the same long drawn-out complicated exchange of electrons, but in the end there was success they became complete and compounds were formed. If you want to see what happened to create Al_2S_3 and B_2O_3 look at the diagrams for making B_2S_3 and Al_2O_3.that we just finished. They are all formed with the same exchange of electrons to become complete.

****

Professor Terry and Guy had been watching all these amazing exchanges of electrons, invisible from the security of their bubble. Professor Terry turned to Guy and said, "Again this is basically what happened as Aluminum and Boron became these compounds. Again you will find out more about covalent bonding in high school. This is the last of the story about how sad little atoms become Happy Atoms. You know the basic secret of compound formation. Tell me."

Guy spoke up and said, "I know, it's getting each atom's Outside Energy Levels complete. The last elements we observed sure had a long drawn out method of accomplishing this."

Professor Terry said, "This was a very involved way to form compounds. There was even much more to it but that's the exciting part of chemistry. There's always more to learn. Remember, I promised to teach you the basics of Chemistry. What I'm teaching you makes it easier to understand the more advanced chemistry they teach in high school. But now it's time to go home." With the magic wand powering the bubble, they were back in the lab in no time.

PROFESSOR TERRY'S LAB AT THE UNIVERSITY

Professor Terry had an experiment to finish and Guy waited around to watch her. She explained, "The chemicals in my erlenmeyer flask are engaging in a slow acting chemical reaction. I was asked to test the liquid to see if it contained what it was supposed to be. Guy, they call this type of experiment Qualitative Analysis. My report of the results is due tomorrow. So I need to finish this experiment before I go home."

While Professor Terry was involved with her work Guy looked around the lab. Of course his attention was drawn to the very large Periodic Table on the back wall. He recalled the first day he arrived at the lab. Wish Star had talked about the Periodic Table but the one in the back of the lab was the first one he had ever seen. He remembered thinking, Is that the Table Wish Star was talking about? It was. Then it turned out that it was far more than an ordinary Periodic Table. It concealed the hidden entrance to Periodic Table Land. It contained the mirrored slide dotted with a thousand sparkling lights that had brought him to Periodic Table Land so often. Guy stood there and looked. There was not a clue that this Table was anything but an ordinary Periodic Table. Yet it was far more. It was the the reason his summer was so special. It was the reason he was learning chemistry first hand from the elements themselves in that mystical world of Periodic Table Land.

Just then Wish Star came flying in and landed on the end of the Lab Table. Wish Star said, “I’ve been keeping up with your progress, Guy. I know you are deep into understanding chemistry now. You have learned all about the Periodic Table. You have drawn Bohr models of the elements. You have even learned how compounds are formed.”

Guy added, “I even know how to make formulas for the compounds.”

Professor Terry added, “And there’s a lot more to come, Guy. There’s still enough time before vacation is over and you have to go back to the city with your parents. You are going to learn about Chemical Reactions, Writing Equations and Balancing Equations. Best of all, you are well prepared to understand these topics.”

Wish Star said, “I’m so proud of you Guy. You have learned so much.”

Professor Terry invited them to look at her experiment. It was a blue liquid and when the reaction completed there was an orange colored metal in the bottom of the flask. “This is what I was waiting for. I was testing this liquid to see if it really contained copper. That’s what you are now looking at in the bottom of this flask. My assignment is over, I can go home now.”

After visiting with them for a while exchanging stories, Wish Star, flew off. Guy left and made his way up the mountain to his parent’s cabin. Professor Terry put out the lab lights after seeing that everything was in order, and she went home, too.

As Professor Terry left she thought, tomorrow I am going to teach Guy the Valence Method of writing chemical formulas. He will love it.

GUY
and
THE VALENCE METHOD

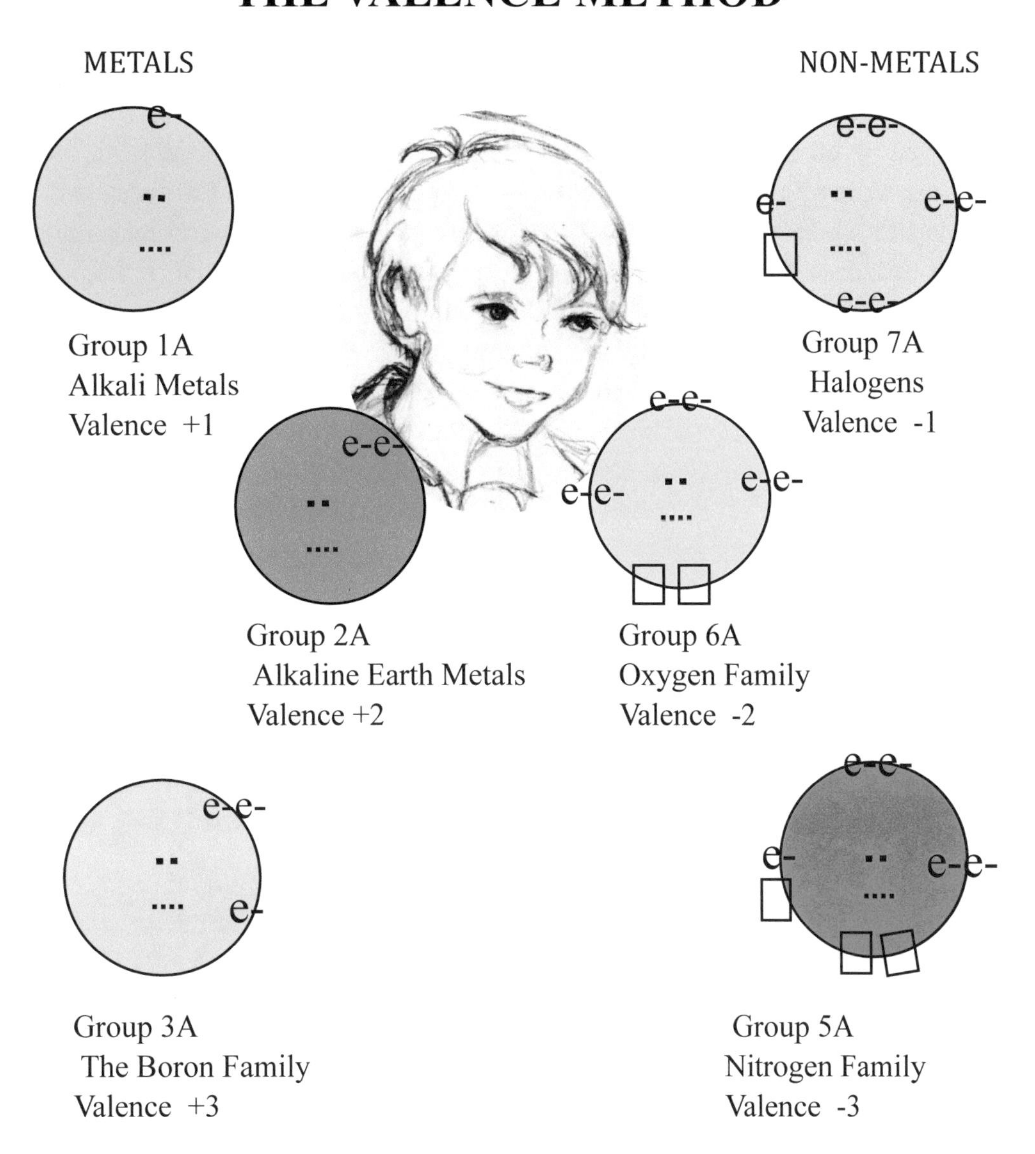

The Valence Method

Ca^{+2} P^{-3}

Ca_3P_2

Guy Learns to Use theValence Method to Write Chemical Formulas
Chapter 11

Summer is coming to an end soon, thought Guy as he got ready extra early to leave for his visit to Professor Terry at the chemistry lab. He needed to talk with her about something that was puzzling him.

Guy entered the lab and saw Professor Terry performing a chemical analysis. Guy loved to watch her at work. So he put on the safety glasses that Professor Terry had given him, climbed up on one of the tall lab stools and sat at a safe distance from the bunsen burner. Here he saw the experiment bubbling away furiously in the Erlenmeyer flask. The liquid was turning to a gas that then flowed through a cooling tube to be collected as a liquid in a beaker at the far end. This is a 'change of state' experiment. It was one of the first things that Wish Star taught me at the beginning of summer. It was fascinating to watch. "Hi Professor, I love seeing you work in the lab. I came to thank you for teaching me how compounds are formed and how to write chemical formulas. But I have a question because the last two compound formations took so long to figure out. I began to think that scientists must have a shorter way to create formulas for compounds. Look what I had to draw here on my pad to be able to discover the formula for the compound formed when Calcium combined with Phosphorus."

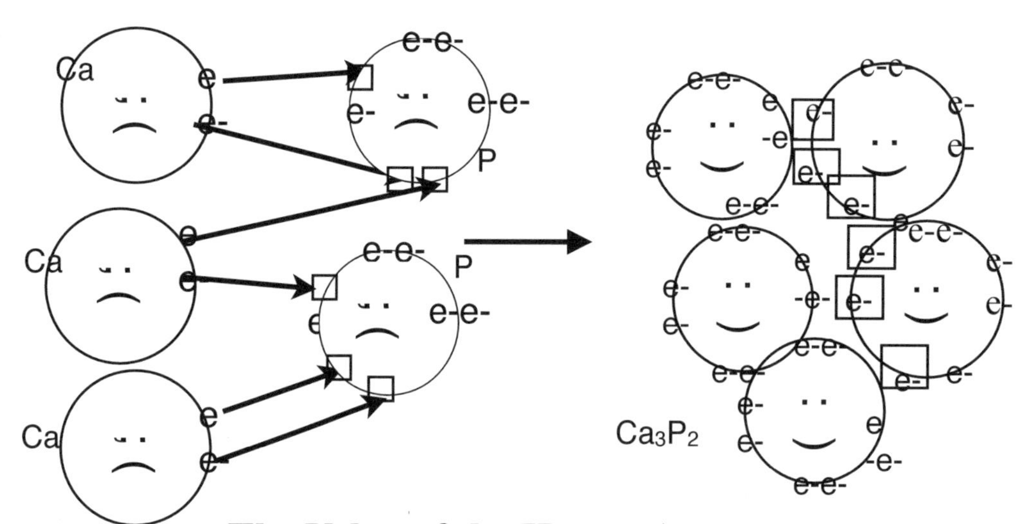

The Value of the Happy Atom Method

Professor Terry's face lit up. "Well Guy, I guess you're ready to learn how scientists construct formulas. The new method of developing chemical formulas is called ***The Valence Method.*** This is the quick way to arrive at the compound's formula.. The Happy Atom story was valuable because it taught you ***why*** the elements were joining together to form a compound. It also helped you understand ***the reason a certain number of atoms of each element were needed to form the compound.*** You learned ***the meaning***

***of the formula*—the symbols stand for the elements that joined to form the compound. You learned that the subscripts in the formulas tell how many atoms of each element were needed to form the compound.** I had to be sure you understood all this before I taught you the Valence Method. TheValence Method gets you the formula for the compound and nothing more. But at this point you understand what the formula stands for. That's what the Happy Atoms taught you. Now let's get started."

Guy responded, "You're right Professor Terry. I am really ready to learn the ***Valence Method***. Finding the formulas for Calcium Phosphide and Aluminum Oxide, made me ready to learn a short method."

Professor Terry agreed and said, "It's time to begin. We will build on what you know as you learn what valence means. It's about atoms getting a charge by either giving away or taking in electrons. Guy, let's see if you can explain how atoms get a charge."

Guy said, "You explained this to me before we began learning about compounds. Let's see if I can remember. When I first learned about the atom I flew around Periodic Table Land and talked to a lot of elements, I found out that the atom of every element has the same number of protons (p+) and electrons (e-). That means an atom has no charge because the + and – charges are equal. An atom gets a charge when it gives away or gets electrons. The number of protons never changes because they are locked up in the nucleus."

"When Sodium and Chlorine form a compound they both get charges. The metal Sodium gives away his 1 electron to the non-metal, Chlorine. Chlorine takes in that 1 electron that Sodium gave away. Both get a complete Outside Energy Level. They both end up with 8 electrons in their Outside Energy Level. Here are the Bohr models to let you see Sodium's 1 e- moving into the empty space in Chlorine 's atom."

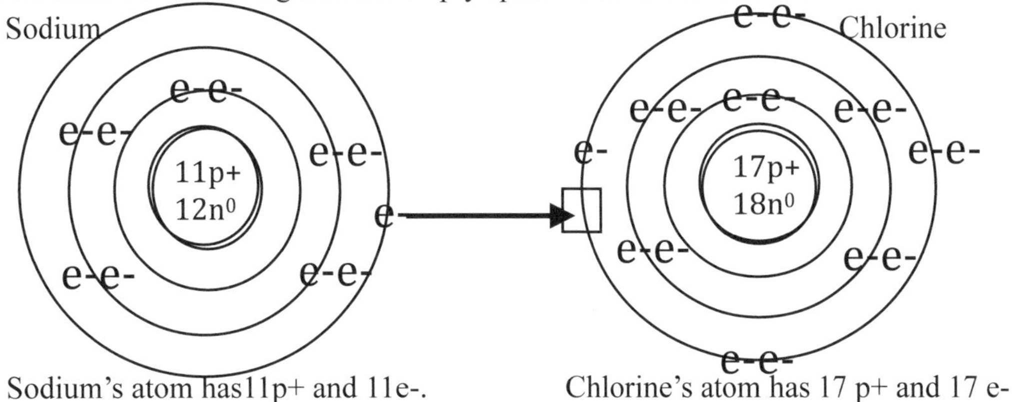

Sodium's atom has11p+ and 11e-.

Chlorine's atom has 17 p+ and 17 e-

Sodium and Chlorine atoms have equal numbers of protons and electrons—no charge.

When Sodium gives Chlorine his 1 Outside electron, they both get charges. Look at **Sodium.** He gave away 1 negative electron to Chlorine. He now is left 11p+ and 10 e-. **He now has 1 more positive proton than negative electrons. His charge is +1**

Look at **Chlorine**. He took in Sodium's 1 negative electron. He now has 18 e-but his protons didn't change. He still has 17 p+. **He now has 1 more negative electron than protons. His charge is -1.**

"Good job Guy! Now I'll tell you the first thing you need to know to understand the valence method. **We call the charge on an atom the atom's valence. Sodium has a valence of +1. Chlorine has a valence of –1. Valence is the same as the charge.**

"Now, I'm going to give you an easy way to think of **valence.** This will make it easier to come up with a formula for a compound when you know which two elements are combining."

"Valence is all about those same electrons in the Outside Energy Level of an atom that you learned about using the Happy Atom method. **Valence is the number of electrons an atom has to give away or get in forming a compound.** I'm sure this sounds familiar. The Metals in Groups 1A have 1 electron to give away. Group 2A metals have 2 electrons to give away and 3A metals have 3 electrons to give away.Their Valences are +1, +2 and +3. The elements in Groups 5A, 6 A, and 7A need to have 8 electrons and be complete. Look how many electrons each Group needs to get to make 8. Group 5A needs 3; Group 6A need 2 and Group 7A needs only 1 more electron to have 8 and have a complete Outside Energy Level. Their Valences are -3, -2, and -1."

"Here is a chart that shows all the Outside Energy Levels of the elements in the A Groups on the Periodic Table and their valences."

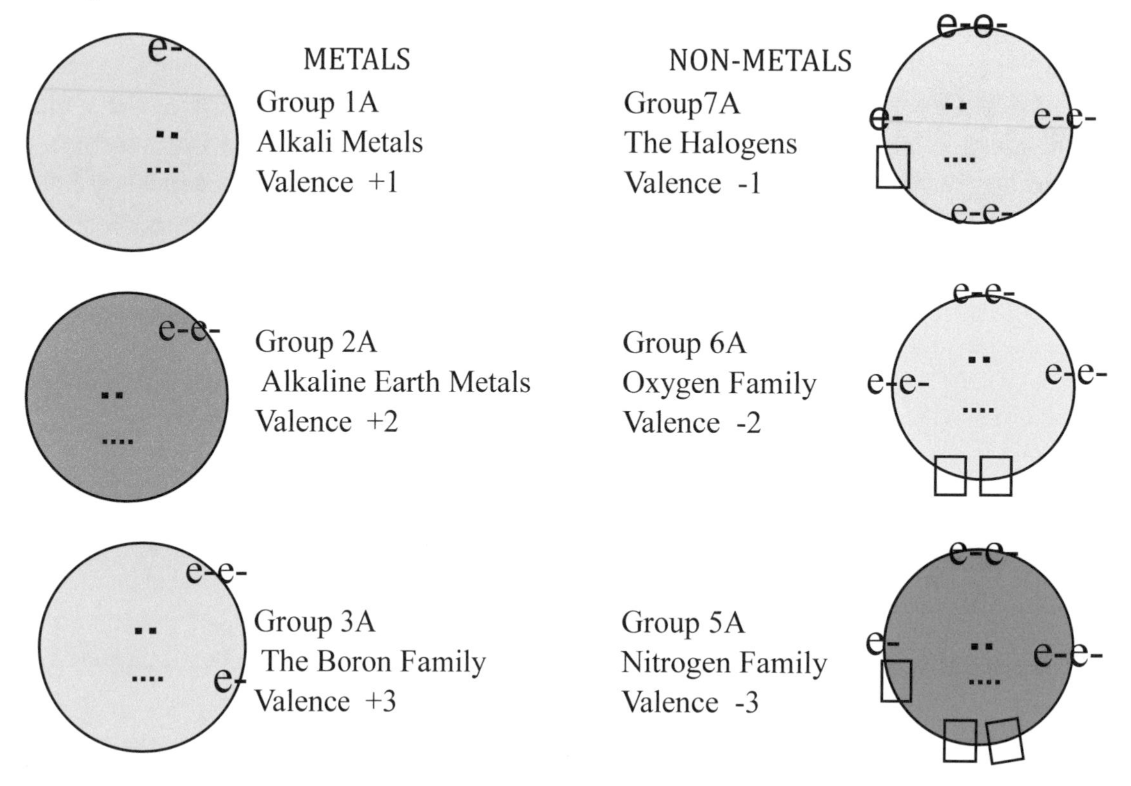

Professor Terry wanted to make sure Guy understood this chart. "Did you notice on the chart the valence is either positive, indicated by the plus sign (+) or negative, indicated by the minus sign (–)."

"Guy, you know that elements naturally have the same number of electrons and protons. So when they give away negative electrons they have more protons and get a positive valence. When they take in negative electrons they add extra negative charges and they get a negative valence."

Metals, Non-metals and Valence

"**The elements in the metal Groups on the Periodic Table** have electrons to give away. So, they have a positive (+) valence, and you know why. When they give away their negative electrons they now have more protons than electrons. This gives them a **Positive Valence**.

The elements in the non-metal Groups on the Periodic Table need to take in electrons to be complete. So they have a negative.(–) valence**,** and you know why. When they take in extra negative electrons they have more negative electrons than protons. This give them a **Negative Valence."**

Compounds and Valence

"The compounds we learned about, using the Happy Atom method were made up of two parts: the metal and the non-metal.

The first part of the compound was the metal–the element that had electrons to give away. Now, we know that part of the compound has a valence that is positive (+). So, the compound begins with the element that has **a positive valence**. The element with the + positive valence is written on the left in the name of the compound and on the left in the formula.

The second part of the compound was the non-metal–the element that was missing electrons and needed to get them. Now we know that part of the compound has a valence that is negative (–). So, the compound ends with the element that has **a negative valence**. The element with the –negative valence is written on the right in the name of the compound and on the right in the formula."

The Groups and Valence

Group 1A elements are called the Alkali Metals. They have 1 electron to give away in forming compounds. That one electron is hiding their Complete Energy Level. When they give away that one negative electron, their complete hidden energy level rises up and becomes their complete Outside Energy Level. They now have one more proton (p +) than electrons. Their valence is +1. Here's what the Outside Energy Level of the atoms in Group 1A looks like.

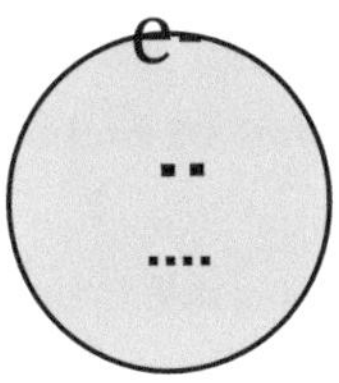

The **Valence is +1 for all the elements in Group 1A**. In forming compounds these elements have one electron to give away. **Remember: Group 1A—valence is +1**

Group 2A elements are called the Alkaline Earth Metals. They have 2 electrons to give away in forming compounds. Those 2 electrons are hiding their Complete Energy Level. When they give away those 2 negative electrons their complete hidden energy level rises up and becomes their complete Outside Energy Level. They now have 2 more protons (p+) than electrons. Their valence is +2. Here's what the Outside Energy Level of the atoms in Group 2A looks like.

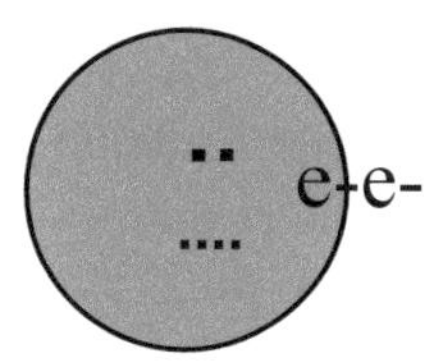

The **Valence is +2 for the elements in Group 2A**. In forming compounds these elements have 2 electrons to give away. **Remember: Group 2A—valence +2**

******

Group 3A elements are called the Boron Family. They have 3 electrons to give away in forming compounds. Those 3 electrons are hiding their Complete Energy Level. When they give away those 3 negative electrons their complete hidden energy level rises up and becomes their complete Outside Energy Level. They now have 3 more protons (p+) than electrons. Their valence is +3. Here's what the Outside Energy Level of the atoms in Group 3A looks like.

e-e-

e-

The **Valence is +3 for the elements in Group 3A.** In forming compounds these elements have 3 electrons to give away. Remember: **Group 3A—valence is +3.**

Remember Valences for these Metal Groups:

Group 1A is **+1**, Group 2A is **+2** , Group 3A is **+3**

******

The Non-metal Groups

"Guy, let me remind you that in Groups 5A, 6A, and 7A the elements are missing electrons. They need to have 8 electrons to get a complete Outside Energy Level when forming a compound. If you think of missing as being minus something, it will be easier to remember that the valence for an element in these groups is minus (–) or negative.

Group 5A is missing 3 electrons to be complete (8 - 5 = 3). The valence is –3.
Group 6A is missing 2 electrons to be complete (8 - 6 = 2). The valence is –2.
Group 7A is missing 1 electron to be complete (8 - 7 = 1). The valence is –1.

Group 5A elements are called the Nitrogen Family. In forming compounds they need to have 8 electrons to have a complete Outside Energy Level. They have only 5 electrons. So they need to take in 3 electrons to be complete. When they take in 3 negative electrons, their valence becomes –3, negative 3. They have 3 more negative electrons than protons. Here's what the Outside Energy Level of atoms in Group 5A looks like.

The **Valence is -3 for all the elements in Group 5A.** This is because they are missing 3 electrons to have a complete Outside Energy Level. The elements need to get 3 electrons to form a compound. **Remember: minus 3 electrons means the valence is –3.**

******

Group 6A elements are called the Oxygen Family**.** In forming compounds they need to have 8 electrons to have a complete Outside Energy Level. They have only 6 electrons. So they need to take in 2 electrons to be complete. When they take in 2 negative electrons, they have 2 more negative electrons than protons. Their valence becomes –2, negative 2. Here's what the Outside Energy Level of atoms in Group 6A looks like.

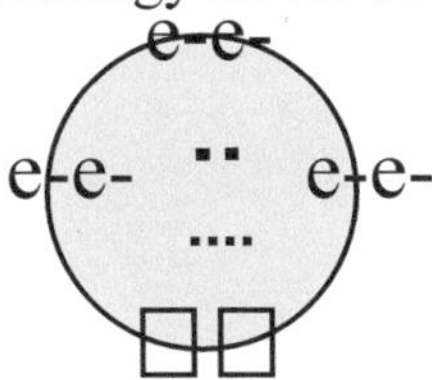

The **Valence is -2 for all the elements in Group 6A.** This is because the elements are missing 2 electrons to have a complete Outside Energy Level. The elements need to get 2 electrons to form a compound. **Remember: minus 2 electrons means valence is –2.**

******

Group 7A elements are called the Halogens. In forming compounds they need to have 8 electrons to have a complete Outside Energy Level. They have only 7 electrons. So they need to take in 1 electron to be complete. When they take in 1 negative electron, they have one more negative electron than protons. Their valence becomes –1, negative 1. Here's what the Outside Energy Level of atoms in Group 7A looks like.

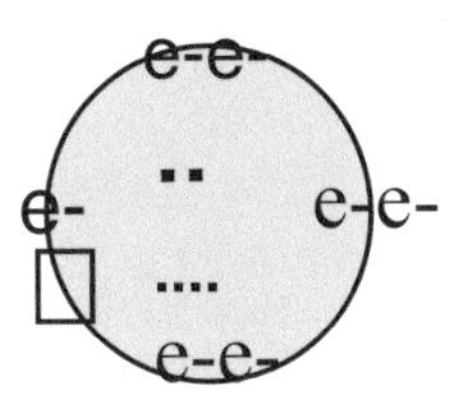

The **Valence is -1 for all the elements in Group 7A.** This is because the elements are missing 1 electron to have a complete Outside Energy Level. **Remember minus 1 electron means valence is –1

Remember Valences for these Non-metal Groups:

Group 5A is -3; Group 6A is -2; Group 7A is -1

The Valence Method

Professor Terry told Guy, "I will teach you the steps that you need to follow to write a chemical formula using the Valence Method. Get out your pad where we created the formula for Calcium combining with Phosphorus using the Happy Atom method."

Guy found it and explained, "It was $\mathbf{Ca_3P_2}$, because it took 3 Calcium atoms to give Phosphorus enough electrons to fill the missing spaces in Phosphorus' Outside Energy Level. You can count them. 3 Ca atoms and 2 P atoms. Remember that allowed Calcium to give away all his outside electrons, enabling his hidden, complete energy level to become his Outside Energy Level."

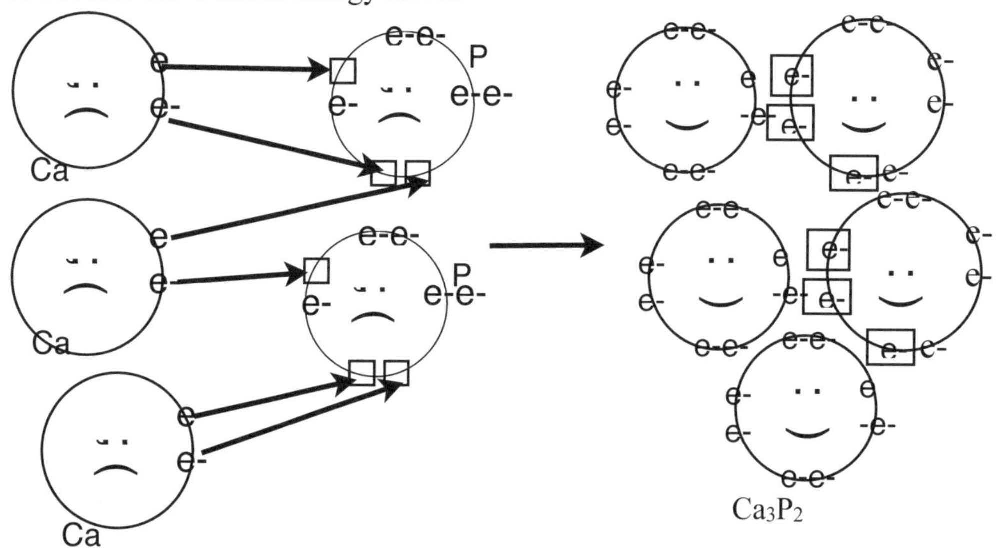

Just then Calcium and Phosphorus popped in with a poster board displaying Calcium Phosphide's formula. We're going to show you the Valence Method can get you this very same formula in a much easier way.

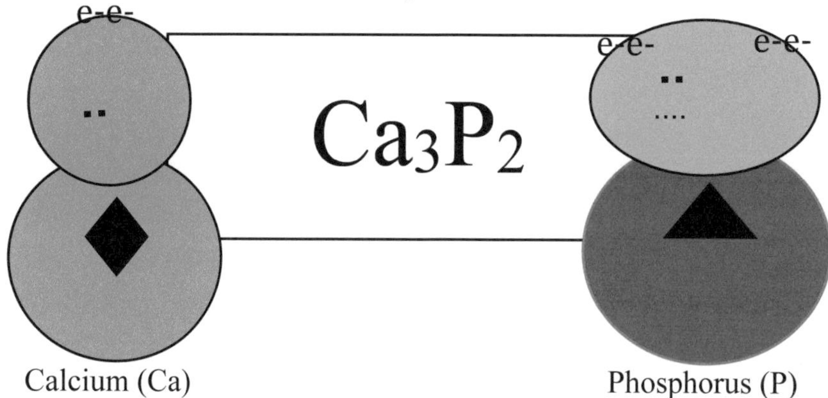

Professor Terry took it and taped it up on the wall saying, "We'll keep this formula for Calcium Phosphide where we can see it, and remember how we arrived at that formula." Professor Terry continued, "Now you are going to learn how to come up with the formula for Calcium and Phosphorus joining using the Valence Method. Here's the chart and Calcium and Phosphorus will show you how it works."

The Valence Method

1.Write the elements to combine

Calcium with **Phosphorus**

2. Write their symbols:

Ca P

3. Write the Group they are in:

Ca is in Group 2A

P is in Group 5A

4. Write their Valence

Ca in Group 2A is + 2

P in Group 5A is – 3

5. Write the Valence as Superscripts

Calcium Ca^{+2}

Phosphorus P^{-3}

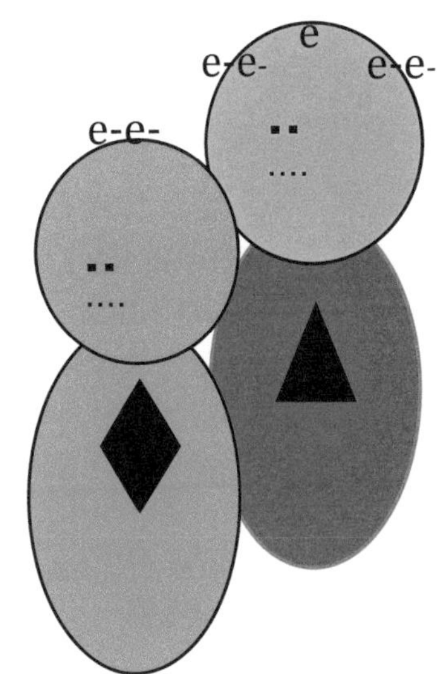

6. Write the two symbols with superscripts together

(Make sure the **(+)** symbol is on the left; the **(–) s**ymbol is on the right)

$Ca^{+2}P^{-3}$

7. Next step is called **"Criss-cross".** Write Calcium's superscript which was +2 as Phosphorus' subscript, P_2 and also Phosphorus' superscript which was – 3 as Calcium's subscript,Ca_3

Notice when I wrote the subscripts. I did not write the + or – sign.

8.Next write the symbols together with the subscripts making sure you write the one that had a + Valence first .

So the formula is Ca_3P_2

Now see if Rule A, B, or C below applies.

Rule A. If any symbol has the subscript 1 Remember we never write ones. Just write the symbol without the subscript 1. If the symbol is there it means there was just one atom of that element. **Rewrite the formula without the 1**.

Rule B. If both numbers are the same. Do not write ANY subscripts. Rewrite the formula with no subscripts.You might remember, if the metal has the same number of electrons that the non-metal is missing, it only takes one atom of each element to combine with each other and we never write ones. We just write the 2 symbols together with no subscripts and that is the formula.

Rule C. If the subscripts are different (and not 1) you have the formula. You are finished!!!

9. The subscripts for both elements in Ca_3 P_2 are different.

10. So Rule C applies. THE FINAL FORMULA IS Ca_3 P_2

Guy checked and said, "The formula came out the same using both the Valence and Happy Atom methods." Guy thanked Calcium and Phosphorus for showing him the steps and rules for writing formulas using the Valence Method.

Then, Professor Terry took over. "Is there any other formula you want to use to compare the Happy Atom Method to the Valence Method?

The Valence and Happy Atom Method Compared

Guy said, "Using the **Happy Atom Method** the formulas for Sodium&Chlorine, Calcium&Oxygen, and Boron&Nitrogen were just the 2 symbols written together with no subscripts: **NaCl, CaO,** and **BN.** This was because these elements joining needed to get the same number of electrons that the other element had to give away? Let's see what the formulas will be using the Valence Method."

The Valence Method

1. **Write the Elements combining**

Sodium & Chlorine,	**Calcium & Oxygen,**	**Boron & Nitrogen**

2. **Write their symbol**

Na & Cl	**Ca & O**	**B & N**

3. **Write Group Number**

1A & 7A	2A & 6A	3A & 5A

4. **Write Valences**

+1 & –1	**+2 & –2**	**+3 & –3**

5. Write **Valence as Superscript**

$Na^{+1} Cl^{-1}$	$Ca^{+2} O^{-2}$	$B^{+3} N^{-3}$

6. Symbols together with the + symbol on the left; the – symbol on right

$\mathbf{Na^{+1} Cl^{-1}}$	$Ca^{+2} O^{-2}$	$B^{3} N^{-3}$

7. **Criss-cross Valences** (leave out + and -)

Na_1Cl_1	Ca_2O_2	B_3N_3

8. Look at the Rules and Decide which rule applies
9. **RULE B**: If Both subscripts are the same. Don't write any subscripts

10. **Final Formula** **Na Cl** **CaO** **BN**

The Valence Method and the Happy Atom Method Compared

Guy said, **"**I'm convinced creating the formulas either way the final formulas come out the same**.** Thanks for teaching me The Valence Method. It's so much faster than drawing all those sad and happy faces, but there are a lot of steps."

Professor Terry wanted to assure Guy that it was even shorter than it appeared. "At first it might seem like a lot of steps, but after a while you can just look up the Group Number and know the valence. Your mind automatically will go through those long steps, and you eventually will be able to know the formula almost instantly when you know which Group the element is in. You will mentally criss-cross the valences to get the subscripts where they should be. Then you will see which rule applies: Like A Subscripts one? Write no ones; Like B Subscripts both the same? Write no subscripts; Like C Subscripts different? That's the formula, and you're done."

Guy said, "When you say it like that, it sounds even shorter."

"Here is the Periodic Table. I know you have another copy but keep this one with the Valence papers. On here, you can find any element and learn which Group it is in. If it's in an A Group, you will know the Valence. If it's in a B Group, you will have to be told what the Valence is. After you use the valences a few times you will remember some of the most common valences." Professor Terry slipped the paper into Guy's folder.

Guy was delighted that he could now write a formula like a real scientist.

Professor Terry said to Guy, "Here are examples of forming compounds using the Valence Method. See if you can tell me what rule was followed to get the final formula. Guy, did you notice I've already made those 9 steps look shorter? Actually I have followed all the steps but just simplified them. You can look at the chart I gave you and see I did follow it completely. The Valence Method is not as long a process as those steps appeared to be. Here's the short version of the 9 Steps in the Valence Method. Your job is to tell me which Rule applies in deciding on the final formula. It could be more than one rule."

Valence Method Shortened

1.Name of Elements combining:
2.Write their symbols:
3.Write their Group #:
4.Write their valence:
5. Name of Compound
6. Symbols with Valences as Superscripts
7. Criss-cross Symbols' valences as Subscripts
8. Final Formula 9. RULE ___

RULES

A. Never write ones. Rewrite formula without the ones.
B. If both subscripts are the same, rewrite the formula without any subscripts
C. If the two subscripts are different and not ones. This is the formula.

Turning to Guy, Professor Terry said, "Below, I've followed these steps and showed you how each of the formulas was created using the Valence Method. I want you to follow each step as if you had to do each step yourself. Then on the line provided, tell me which Rule or Rules you would use to get the final formula."

Quiz : Which Rule Applies, A, B, or C ?

1.	Potassium	Fluorine	Potassium Fluoride	
Symbol	K	F	$K^{+1} F^{-1}$	
Group	1A	7A	$K_1 F_1$	Rule ______
Valence	+1	–1	KF	

2.	Magnesium	Sulfur	Magnesium Sulfide	
Symbol	Mg	S	$Mg^{+2} S^{-2}$	
Group	2A	6A	Mg_2S_2	Rule _____
Valence	+2	-2	MgS	

3.	Calcium	Nitrogen	Calcium Nitride		
Symbol	Ca	N	Ca^{+2} N^{-3}		
Group	2A	5A	Ca_3N_2	Rule	___
Valence	+2	-3	Ca_3N_2		

4.	Sodium	Oxygen	Sodium Oxide		
Symbol	Na	O	$Na^{+1}O^{-2}$		
Group	1A	6A	Na_2O_1	Rule	___
Valence	+1	-2	Na_2O		

5.	Lithium	Phosphorus	Lithium Phosphide		
Symbol	Li	P	$Li^{+1}P^{-3}$		
Group	1A	5A	Li_3 P_1	Rule	___
Valence	+1	-3	Li_3 P		

What rule was used ? Answers 1. A & B, 2. B, 3. C, 4. A, 5. A,

A Valence Method Challenge for Guy

"OK Guy! This is it. Can you create a formula for a compound using the Valence Method? You have all the helps above. Create the compound made with the following two elements: Magnesium and Nitrogen. Do this and you will be able to say you can write formulas using the Valence Method."

At that moment in popped Magnesium and Nitrogen saying, "We're here to help you." Guy was so excited to see them.

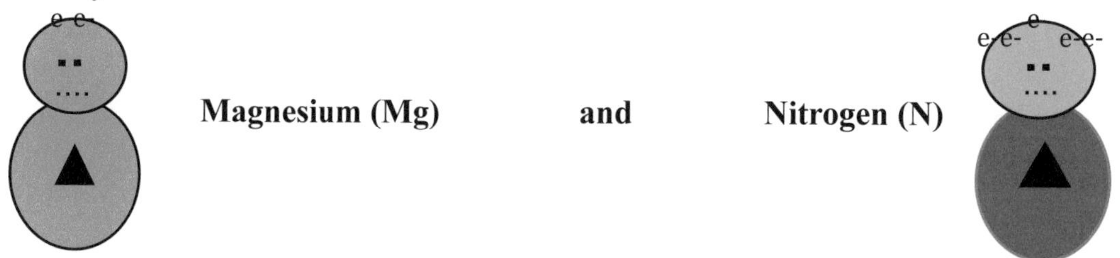

Professor Terry continued, "I'll give you a little help too. Use the set up below. It's a short version of the steps 1 to 9 above: You can look there to see how to do it. Good Luck, Guy!"

Write Elements combining:	Name of Compound
Write their symbols:	Symbols with Valences as Superscripts
Write their Group #:	Criss-cross Symbols' valences as Subscripts
Write their valence:	RULE ___ Final Formula_______________

Guy began. Magnesium and Nitrogen watched because Guy didn't really need help.

Elements	**Magnesium**	**Nitrogen**	Magnesium Nitride
Symbols	Mg	N	Mg^{+2} N^{-3}
Group	2A	5A	Mg_3 N_2
Valence	+2	-3	RULE C Mg_3N_2

"You did that very well, Guy. I think you're ready for one last challenge. See if you can write a formula given the name of the compound. Here's the compound:

Sodium Sulfide.

"I'll give you a hint. First you have to figure out and write the elements that were combining to make this compound."

Guy said, "I know that compound is made up of the two elements because it ends in ***ide.*** So the two elements are **Sodium and Sulfur."**

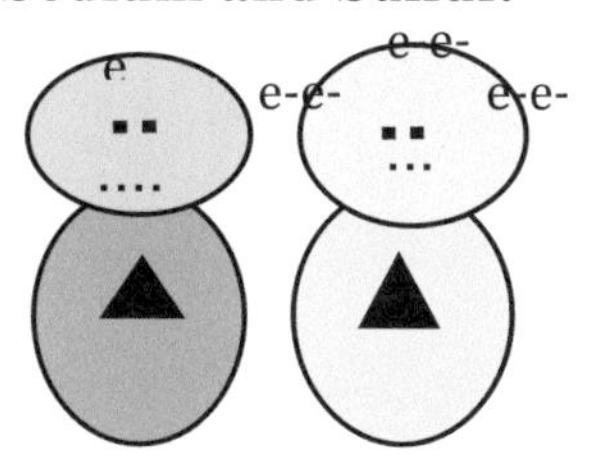

Guy looked at the short version of the valence method steps and began to figure out the formula for the compound when Sodium and Sulfur combine. Guy said, "I have the name of the compound. So, I'll write that in first. Then I'll write the elements that made that compound—Sodium and Sulfur." That was a start. With the rules in front of him, he kept following the rules until he was finished.

Rules

Write Elements combining.	Name of Compound
Write their symbols:	Symbols with Valences as Superscripts
Write their Group #:	Criss-cross Symbols' valences as Subscripts
Write their valence:	RULE ___ Final Formula______________

Elements:	Sodium	Sulfur	Sodium Sulf***<u>ide</u>***
Symbols:	Na	S	Na^{+1} S^{-2}
Group #:	1A	6A	Na_2S_1
Valence:	+1	-2	RULE A Na_2S

"Well Guy, you did it. You know how compounds are formed. You even know how to write chemical formulas. Now you're ready to learn Chemical Reactions. That's your next challenge in chemistry. *Chemical Reactions* is the next topic that will help you learn more about our magnificent world. We will be going off to Periodic Table Land again. Our job this next time will be to show the elements another way to become even happier Happy Atoms."

Guy was delighted at the thought of returning to Periodic Table Land. He went home thinking up all sorts of fun this could turn out to be before his vacation ended on Labor Day.

Wish Star and Guy on the Mountain

Before he went to bed, Guy slipped out the back door and made his way along that familiar path that led to his favorite place on his mountain. He sat and rested his head on his favorite rock searching the star studded sky for his favorite constellations. It was so peaceful and the aroma of honeysuckle and pine always made him happy. It was real quiet until suddenly, he heard a sound coming from the top of a nearby tree. Looking up to see if the wind had kicked up, Guy saw..guess who?

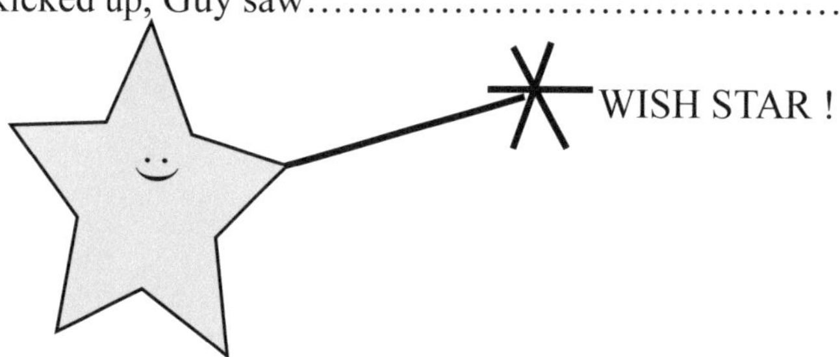

It was Wish Star. Guy was so very happy to see his dear friend, Wish Star."I haven't seen you for a while Wish Star, and I've missed you. I have learned so much since we met here on the mountain my first night of vacation. I can even write chemical formulas for compounds now using the Valence Method."

Wish Star commented, "That's a big leap from that first day when I introduced you to Professor Terry and she taught you the parts of an atom. Guy, I'm so very happy to see you again. I've been watching you from the sky. However it's so much more fun to sit here on the mountain and visit with you." They sat and chatted for a long time—laughed and exchanged stories. They spotted so many familiar constellations. Wish Star knew them all. Occasionally they would catch sight of a shooting star. Guy loved the night sky.

Wish Star said, "I heard you'll be learning about chemical reactions next. You'll love it Guy and I'll be back to check on you. You can count on it. Off he flew high into the night sky. As always he turned one last time and waved."

Guy sat quietly for a time thinking of all the fun he was having this summer. Magic everywhere! He thought of the silly electrons who sang their greeting to him. Professor Terry's Periodic Table with the mirrored slide that took him to Periodic Table Land. He recalled Mr. Mischief who didn't want to put his play dough proton in the pretend nucleus. Guy thought, "Who spends their summer flying around a fantasy world in a magic bubble? I am so lucky. What a wonderful summer I'm having! I really can't decide what I liked best and I don't have to decide. I like every part of my adventure. I'm just having a great time and there's still more time left for lots more fun. Tomorrow I'm going back to Periodic Table Land again. With that thought he made his way back to the cabin. He couldn't wait for tomorrow's adventure. When Guy got into bed, he drifted off to dreamland so very happy.

Wish Star says, "You've read **BOOK 3.** You know how **Compounds** are formed. You learned how to write **Chemical Formulas**. You even learned the **Valence Method**

Now, it's time to read **BOOK 4.** Learn about the **Polyatomic Ions** and their compounds. Study **Chemical Reactions.** Learn to write and balance **Chemical Equations.** *Enjoy the. Labor Day Parade that celebrates Guy's summer Chemistry Adventure.*

***The Happy Atom Story* is**

MAGICAL !

Printed in the United States
By Bookmasters